Armin Täubner

Phant
plastis

Dreidimensionale Figuren aus Tonpapier

frechverlag

Von dem bekannten Autor Armin Täubner sind im frechverlag zahlreiche andere Titel erschienen. Hier eine Auswahl:

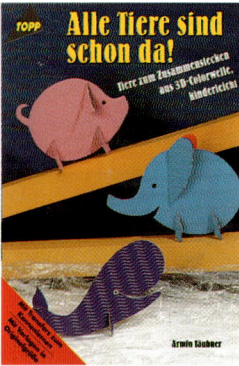

TOPP 2127

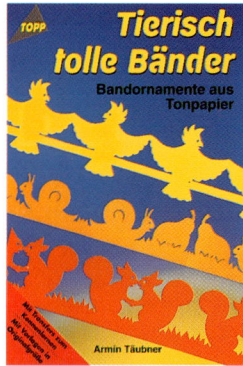

TOPP 2130

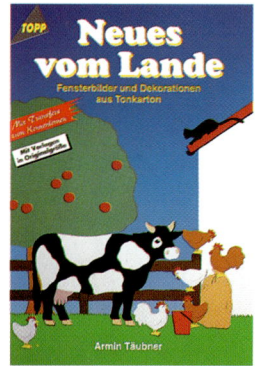

TOPP 2094

TOPP 1993

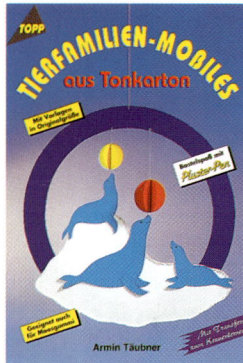

TOPP 1917

TOPP 2304

Fotos: frechverlag GmbH + Co. Druck KG, 70499 Stuttgart; Birgitt Gutermuth

Dieses Buch enthält: 1 Vorlagenbogen

Auflage: 8. 7. 6. 5.
Jahr: 2002 01 00 1999 | Letzte Zahlen maßgebend

© 1996

ISBN 3-7724-2129-6 · Best.-Nr. 2129

frechverlag GmbH + Co. Druck KG, 70499 Stuttgart

Druck: frechverlag GmbH + Co. Druck KG, 70499 Stuttgart

Papier und Klebstoff sind bekannte Bastelutensi-lien. Aber so richtig phantastisch plastisch wird´s erst, wenn neben Tonpapier und Schere auch eine Nähmaschine ins Spiel kommt!

Auf die Grundform einer Figur – zum Beispiel eines Tieres – werden immer mehrere Tonpapier-Körperteile aufgenäht, die zuvor deckungsgleich gearbeitet worden sind. So entstehen dreidimensio-nale Fische, Blüten, Enten und vieles mehr. Alternativ zur Nähmaschine können Sie die Papierschichten natürlich auch per Hand zusam-mennähen oder – für ganz Schnelle – mit dem Klammerapparat zusammenheften.

Die Küken bekommen ein gelbes Federkleid, die Schnecken tragen die schönsten Häuser durch die Landschaft. Selbst der Bauch der Eule wird mit die-ser Technik kugelrund, und das schneeweiße Kleid des Engelchens zeigt sich hier besonders hübsch. Die Kaktus-Stacheln sehen aus, als ob man sich daran stechen könnte, und bei der vollfruchtigen Birne möchte man am liebsten reinbeißen ...!

Sie sehen, phantastisch plastisch ist eine tolle Sache, die auch kleineren Kindern Spaß macht!

So wird´s phantastisch plastisch
Schritt-für-Schritt-Anleitung ▬

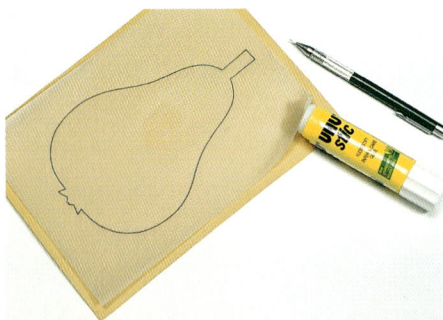

Pausen Sie Ihr ausgewähltes Motiv auf Transparentpapier ab, das Sie anschließend auf einen dünnen Karton kleben.

Nun wird die Schablone mit der Schere oder dem Cutter ausgeschnitten.

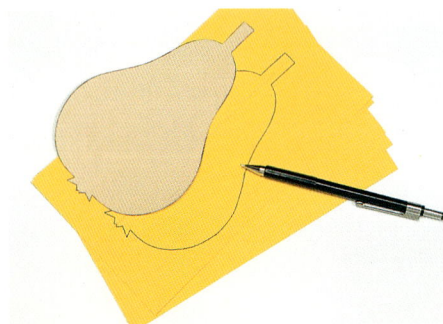

Stapeln Sie mehrere Tonpapierstücke, die etwas größer als die Schablone sind, aufeinander. Auf das oberste übertragen Sie mit Bleistift den Umriß der Schablone; auch den Verlauf der späteren Naht (Strich-Punkt-Linie) deuten Sie mit einem dünnen Bleistiftstrich an.

Damit der Papierstapel nicht verrutscht, wird er außerhalb des Motivs mit dem Tacker zusammengeheftet. Nun können Sie das Motiv in einem Arbeitsgang mehrmals ausschneiden.

Alle ausgeschnittenen Teile werden in der benötigten Anzahl exakt aufeinander gestapelt und mit der Nähmaschine zusammengenäht. Hierbei ist es hilfreich, wenn Sie die Tonpapierteile mit Büroklammern zusammenhalten.

Wer keine Nähmaschine besitzt, kann entlang der späteren Naht (dünne Bleistiftlinie) mit einer Vorstechnadel im Abstand von 3 bis 5 mm kleine Löcher einstechen und später Nadel und Faden durchziehen. Damit die Teile dabei nicht verrutschen, sollten sie beispielsweise mit Wäscheklammern oder Büroklammern zusammengehalten werden. Eine weitere Alternative ist der Tacker. Die Klammern müssen dabei allerdings genau auf der Nahtlinie sitzen, sonst läßt sich die fertige Figur später nur schlecht aufklappen. Innenlinien oder Augen werden übrigens mit einem schwarzen Filzstift aufgemalt.

Besetzt!

Für jeden Apfel werden vier deckungsgleiche Teile aufeinander genäht; anschließend können noch die zweiteiligen Bewohner aufgeklebt werden, deren freundlich lachendes Gesicht aufgemalt wurde.

Entenpaar

(Beschreibung siehe Seite 8)

7

Entenpaar (Abbildung siehe Seite 7)

Legen Sie auf das braune Kopf-Rumpf-Teil (durchgezogene Linie) das große, weiße Rumpfteil (langgestrichelte Linie) und darauf zwei kleine, weiße Rumpfteile (kurzgestrichelte Linie). Diese vier Teile werden entlang der Strich-Punkt-Linie zusammengenäht. Klappen Sie jeweils das braune Kopf-Rumpf-Teil nach hinten, und kleben Sie die beiden Köpfe zusammen. Dann bekommt jedes Tier noch seinen Schnabel und das Auge – und zwar möglichst auf beiden Körperseiten.

Panzertiere

Auf das Beinteil (langgestrichelte Linie) legen Sie zwei Panzerteile (kurzgestrichelte Linie) und darauf ein Kopf-Körper-Teil (durchgezogene Linie). Nähen Sie diese vier Teile an der Strich-Punkt-Linie zusammen. Dann werden die beiden Kopfteile aufeinander geklebt. Und das Panzertier braucht auf beiden Seiten noch sein Auge.

Frisch geschlüpft

Die Eier bestehen jeweils aus acht aufeinandergenähten Teilen; die Beschreibung des Kükens finden Sie auf Seite 12.

Schwanen-See

(Beschreibung siehe Seite 12)

Schwanen-See

(Abbildung siehe Seite 11)

Auf das Kopf-Rumpf-Teil (durchgezogene Linie) werden das große Rumpfteil (langgestrichelte Linie), das mittlere und das kleine Rumpfteil (kurzgestrichelte Linie und punktierte Linie) genäht. Anschließend kleben Sie die beiden Köpfe aufeinander, bringen auf beiden Körperseiten den Schnabel an und malen das Auge auf.

Kleine Federbällchen

Auf das Kopf-Rumpf-Teil (durchgezogene Linie) nähen Sie auf beiden Seiten jeweils drei Rumpfteile (gestrichelte Linie). Der Schnabel wird in der Mitte gefaltet (gestrichelte Linie). Kleben Sie jeweils die untere Schnabelhälfte zusammen mit den Augen auf den Kükenkopf.
Alternativ können Sie auch ein Küken ohne Flügel herstellen: Hier wird das Kopf-Rumpf-Teil einfach ohne Flügel gearbeitet.

Stachelige Gesellen

Nähen Sie fünf gezackte Rumpfteile entlang der Strich-Punkt-Linie aufeinander. Bis auf das unterste Teil werden alle Stachelteile nach oben geklappt, so daß ein hübsches Stachelkleid entsteht.

Das halbkreisförmige Kopfteil wird an den beiden gestrichelten Linien umgeklappt. Schieben Sie das Kopfteil so auf den zusammengenähten Igelrumpf, daß sich jeweils der Anfang einer Stachelreihe in einem Einschnitt des Kopfteils befindet. Die beiden abgeknickten Seiten des Kopfteils werden an der Unterseite des Rumpfes angeklebt.

Abschließend bringen Sie noch die Augen der stacheligen Gesellen an.

Prächtige Häuser

Am gelben Rumpf-Haus-Teil (durchgezogene Linie) werden beidseitig jeweils ein lilafarbenes und ein gelbes Hausteil (gestrichelte Linie) aufgenäht. Und wer mit solch einem prächtigen Haus durch die Gegend wandert, braucht noch auf beiden Körperseiten ein aufgemaltes Gesicht.

Mümmelmänner

Nähen Sie auf das Kopf-Rumpf-Teil (durchgezogene Linie) auf Vorder- und Rückseite jeweils drei Rumpfteile (gestrichelte Linie). Dann noch das Gesicht aufmalen – fertig ist der Hase!

Unter Wasser

Beidseitig werden jeweils drei lilafarbene Rumpfteile (gestrichelte Linie) auf das gelbe Kopf-Rumpf-Teil (durchgezogene Linie) genäht. Augen nicht vergessen!

Eulen-Trio

Ergänzen Sie zuerst den Schnabel und anschließend die Augen auf dem Eulenkopf. Auf Vorder- und Rückseite des Kopf-Rumpf-Teils (durchgezogene Linie) werden jeweils drei Rumpfteile (gestrichelte Linie) genäht.

In fröhlicher Runde

Am Kopf-Rumpf-Teil (durchgezogene Linie) werden zuerst die Vorder- und Hinterbeine an der gestrichelten Linie umgeklappt. Anschließend ergänzen Sie die Augen und malen das Froschgesicht auf. Nun am Kopf-Rumpf-Teil auf Vorder- und Rückseite jeweils zwei Rumpfteile (gestrichelte Linie) aufnähen.

Schmetterlings-Treff

(Beschreibung siehe Seite 22)

Schmetterlings-Treff

(Abbildung siehe Seite 21)

Das rote Kopf-Flügel-Teil des Schmetterlings wird beidseitig jeweils mit drei gelben Rumpfteilen (gestrichelte Linie) geschmückt. Und natürlich dürfen auch die Flügelpunkte und die Augen nicht fehlen.

Wie im Schlaraffenland

Nähen Sie vier Rumpfteile aufeinander. Am halbkreisförmigen Kopfteil werden die beiden Seiten an den gestrichelten Linien nach unten umgeschlagen und am Rumpf angeklebt. Bringen Sie noch die großen Ohren, die Augen und den Schwanz an.

Glückskäfer

Auf das schwarze Kopf-Rumpf-Teil (durchgezogene Linie) nähen Sie drei rote Flügelteile (gestrichelte Linie). Abschließend werden noch die Augen ergänzt.

Kugelrunde Teddybären

Kleben Sie zuerst das weiße Schnauzenteil auf, und malen Sie dann das Bärengesicht auf. Nun können am Kopf-Rumpf-Teil (durchgezogene Linie) beidseitig jeweils zwei Rumpfteile (gestrichelte Linie) aufgenäht werden.

Blütengrüße

Auf das große Tulpenteil
nähen Sie beidseitig
jeweils zwei Blütenteile.

Kugelrund und rosa

Legen Sie am Rumpfteil 1 beidseitig jeweils einmal Rumpfteil 2 und 3 an. Das Rumpfteil 2 und das Beinteil können alternativ auch als ein Teil ausgeschnitten werden. Nähen Sie die fünf Rumpfteile zusammen; der Nahtverlauf wird dafür auf Rumpfteil 3 mit einer dünnen gestrichelten Linie angedeutet. Die Ohren knicken Sie an der punktierten Linie ab und kleben sie jeweils oben ans Rumpfteil 2. Zum Schluß werden noch die Augen und der Mundwinkel aufgemalt.

Erntezeit

Bei den Birnen werden sechs, bei den Zwetschgen vier gleiche Teile aufeinander genäht.

Stachelkugeln

Die Kaktusteile werden hier mit der Zackenschere ausgeschnitten; nähen Sie jeweils sechs Teile zu einem Kaktus
zusammen. Die Blüten bestehen jeweils aus fünf aufeinandergenähten Teilen.

Schneemänner

Für die Schneemänner werden auf das Kopf-Rumpf-Teil (durchgezogene Linie) beidseitig drei Rumpfteile (gestrichelte Linie) genäht.

Soll der Schneemann stehen, werden sowohl das Kopf-Rumpf-Teil als auch die zweimal drei Rumpfteile unterhalb der Arme anders ausgeschnitten, nämlich entlang der punktierten Linie (siehe Vorlagenbogen). Abschließend ergänzen Sie noch den Hut und malen das Gesicht auf!

Engelschar

(Abbildung siehe Seite 32)

Auf das Kopf-Rumpf-Teil (durchgezogene Linie) nähen Sie beidseitig drei Kleidteile (gestrichelte Linie). Kleben Sie nun das Haarteil samt Gesicht und die Hände an. Die beiden Flügel werden jeweils von hinten befestigt.

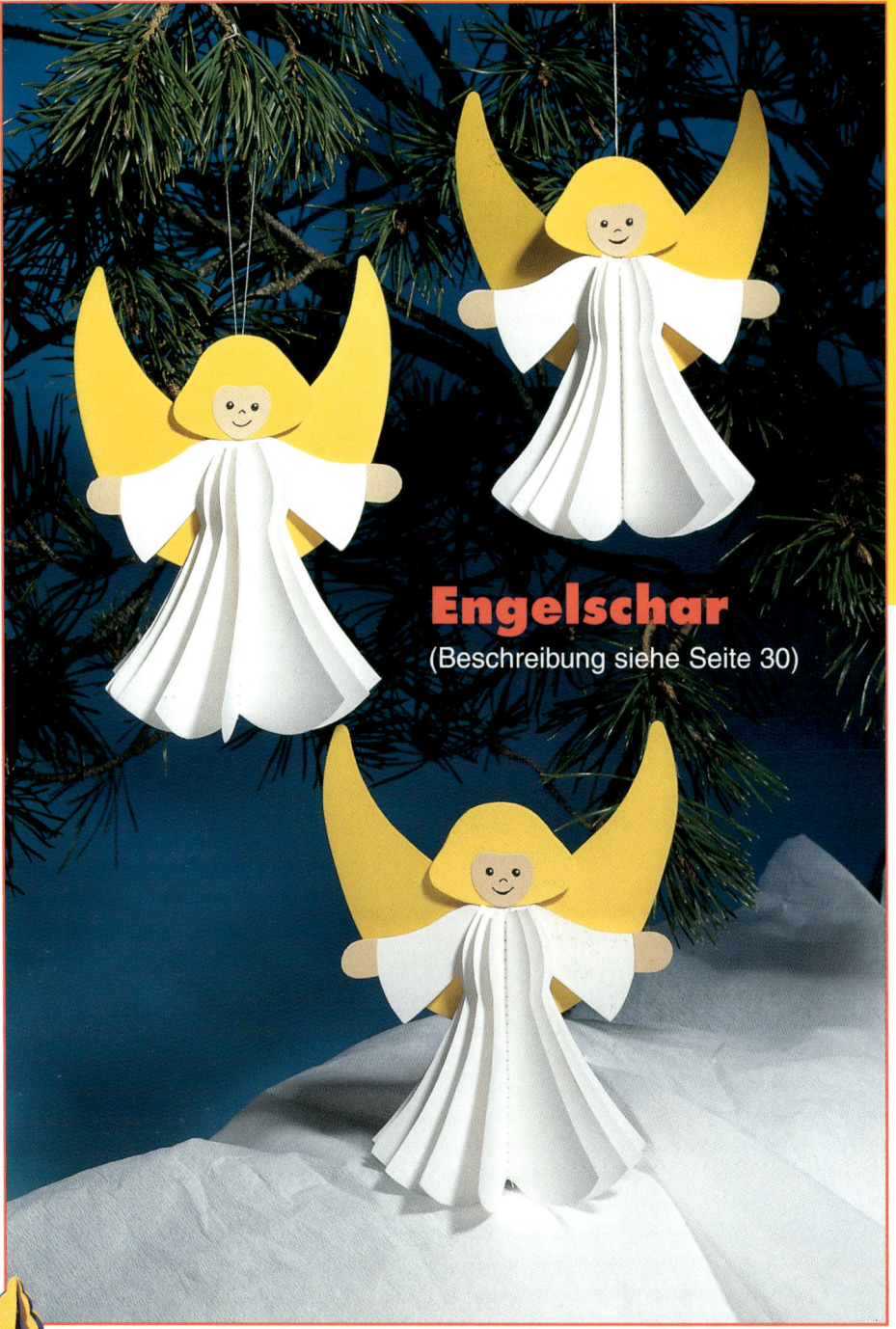

Engelschar

(Beschreibung siehe Seite 30)

OCCULT DETECTIVE
MAGAZINE

#8 DECEMBER 2021

edited by

JOHN LINWOOD GRANT and DAVE BRZESKI

CATHAVEN
PRESS

OCCULT DETECTIVE MAGAZINE #8

ISBN: 978-1-9160212-4-2
http://greydogtales.com/blog/occult-detective-magazine/
occultdetectivemagazine@gmail.com

Publishers: Jilly Paddock & Dave Brzeski
Editors: John Linwood Grant & Dave Brzeski
Logos & Headers: Bob Freeman & Mutartis Boswell

Cover by: Stefan Keller (https://pixabay.com/users/kellepics-4893063/)

Interior design by Dave Brzeski and Jilly Paddock

The Voice on the Moor art © 2021 ODM/Andy Paciorek
(https://www.blurb.co.uk/user/andypaciorek)
The Dead Shall Rise art © 2021 ODM/Mutartis Boswell
(https://boswellart.blogspot.com/)
Memory Fumes art © 2021 ODM/Mutartis Boswell
The Grey Men of Glamaig art © 2021 ODM/Autumn Barlow
(https://issybrooke.com/illustration/)
Vinnie de Soth and the Saucer People art © 2021 ODM/Mutartis Boswell

Published by
Cathaven Press,
Peterborough,
United Kingdom
cathaven.press@cathaven.co.uk

CONTENTS

EDITORIAL

JOHN LINWOOD GRANT

We return! Our usual regular, irregular schedule was interrupted by the release of Occult Detective Magazine #0, our bumper free issue, and our special revised edition of Rosemary Pardoe's excellent collection The Angry Dead, but now we're back to the main magazine with #8. Not only that, but #9 is due early next year, and #10, also coming out in 2022, will highlight longer stories which we've been dying to share, but which couldn't normally be fitted in.

This issue is packed with both fiction and non-fiction – including lots of reviews, an aspect we hope to extend – in fact we have the first in a semi-regular (that is, it'll appear whenever we have something relevant to talk about) column, *Dicing With Death*, which will cover games of some relevance to our sub-genre. Once again we've sought out tales which show the sheer breadth of the sub-genre. If you think 'occult detectives' are simply a few Victorian amateurs, an out-of-luck noir PI in a trench-coat, or a bit of urban fantasy, you must have missed our earlier issues. In these pages, we range across time and continents, from worrying peculiarities to outright horror.

D.G. Laderoute conjures a dark contemporary tale of a serial killer called 'The Piper', and Rebecca Buchanan brings us a Judge's apprentice in a story steeped in the supernatural threats of 1904 Haiti. Cristina L. White muses on Art Deco and transformation, introducing a thoughtful psychic investigator who looks into art mysteries; Rhys Hughes takes us into the surreal world of Svevo, a city polluted by... nostalgia, and Uchechukwu Nwaka provides a glimpse into the life of a Nigerian spirit counsellor, a man with a fearsome horror in his own past.

We also have a historical tale from Carsten Schmitt; a determined Goth necromancer from C.L. Raven; a slice of more-than-haunted house by Jonathon Mast, and a most unusual take on what really lies behind the weird stories and rumours we think we know, by Robert Guffey. Taking an entirely different approach, we can add Dr Johnson and his friend Boswell, who puzzle over a strange phenomenon in the Highlands, in a tale by Andrew Neil MacLeod.

As if these weren't enough, we're delighted to provide new stories featuring return characters. I.A. Watson's Vinnie de Soth (from ODM #6) probes encounters of the third kind – maybe? – in rural England, and Melanie Atherton Allen's Simon Wake, last seen in that same issue, flickers in from the Realm of the Fae to confuse his friends – yet again. Finally, we offer fresh adventures for Paul StJohn Mackintosh's Scottish, bureaucratically-minded Ghost Adjuster, and Brandon Barrows' Japanese priest-investigator Azuma Kuromori, both characters featured in ODM #7.

Dig in, dig deep, and enjoy.

JOIN THE
OCCULT DETECTIVE MAGAZINE
MAILING LIST!

For quite a while now, we've relied on the Occult Detective Magazine Facebook group, and the Doomed Meddlers page on John Linwood Grant's GreyDogTales blog to keep people informed about what's happening with the magazine.. Not everyone is on Facebook however, so we decided an email based mailing list might be a good thing.

The list will contain information about upcoming issues, open calls for author/artists, special projects and the like. We will not spam you with endless emails. In fact, we'll only send an update when we have something specific to pass on. While it's possible that you may get two emails in one week, it's just as likely that you'll go months without seeing one.

To sign up for the mailing list, simply send an email with "Join the ODM Mailing List" as the subject line to occultdetectivemagazine@gmail.com.

THEATRE OF THE MIND

D.G. LADEROUTE

I arrived at the hospital just a little too late. Just like I knew he would, Rob Daniels turned and paced right into the path of a gurney being wheeled past the nurses' station. The thing nearly tipped, threatening to spill a gut-shot girl with a greasy punk hairdo onto the floor. Rob backed off, scowling at the pair of orderlies pushing the girl towards surgery.

"Rob, are you okay?" I called – a needless question, because I knew damned well he'd be okay. Just like I'd known, a good half-hour earlier, that this whole little vignette would happen exactly this way.

He straightened the Toronto police badge dangling from his jacket. "Hi, Jen. Yeah, I'm fine." He raised his voice a notch. "No thanks to the freakin' angels of mercy, though!"

One of the orderlies tossed him the finger, without looking back.

Without thinking, I said, "Sorry, I should have..."

He looked at me, his walrus-face still half-caught in a glare. "Should've what?"

Oops... I shook my head. "Never mind. So, anyway, what's so important you had to drag me out of bed?"

A tired smile lifted his moustache. "You mean you didn't know ahead of time I was going to call?"

I guess my glance was a little too quick, because the smile wavered. I *hadn't* foreseen the call, but his sidelong reference to prescience rattled me. And I wasn't sure why. Prescient episodes hit me often enough, almost always in fore-reference to something inconsequential – like the time I foresaw breaking a vase my mother had given me. It happened, sure enough... a shame, but hardly a disaster.

I covered by smiling back. "Yeah, well, I answered it anyway." So, change the subject again. "So, what's this all about?"

The smile vanished like a light flicked off. "In here," he said, and led me down the hall, through the customary hospital-reek of smells sterile and septic and into one of the rooms.

Two beds. One was occupied.

A boy, maybe ten, maybe twelve years old. Handsome, or he seemed it, anyway, under the bandages. Too thin, though. Malnourished. Scars, old and new. A street kid, probably...

Then it hit me, what this was about.

I looked at Rob. "The Piper, right?"

He nodded.

Shit. First the prescience. Now this.

The Pied Piper was a monster... a brutal serial killer who preyed on Toronto's swollen population of street kids. His nickname, coined by the media, alluded to the fact that no matter how many warnings the cops spun onto the streets, he still managed to somehow lure his young victims to secluded spots, and kill them – eventually.

Eighteen, so far, boys and girls both. And – as far as the news reported, anyway – there were no leads, no physical evidence, not even anything useful from other paranormals involved in the case.

"He was staggering down Danforth," Rob said, "when he got swiped by a car. Just another doped-up street kid, the paramedics thought. Then they put away their guns and started looking close at his injuries." He ran a hand through thinning hair. "It fits the M.O., which means either the Piper screwed up, or he was interrupted. Hopefully, the latter, because we're looking for witnesses right now. In the meantime, though..." He looked at me.

I looked away.

Ordinarily, I'd jump at a high-profile case like this. But I'd never worked with kids before. That was a whole empathic field on its own. More than all that, though, I just didn't know children. The way my life had evolved, kids just hadn't ever figured into it. Somehow, the thought of bringing them into it now, in the person of this boy... well, the idea made my stomach tighten down to nearly nothing.

I shook my head. "Look, Rob... you should try Rachel Chin. She's the best empath I know, and she's worked with children—"

"We already did. She's in Ottawa, tied up with the Supreme Court, expert witness, or something. And, before you ask, no other paras are available, either."

I searched his face instinctively, but there was no judgement there. That was one of the things I liked about Rob. To him, paranormality was just another part of the business – another investigative tool, like forensic pathology or ballistics or DNA typing. That attitude was rare at best, and not just among cops. Almost a generation had passed since the genome folks had unravelled the genetic basis for paranormality, making it something scientific and understandable. But most people nonetheless put paranormality on par with voodoo and seances – except scarier, because it really worked and never mind all the limitations and controlled conditions required. Add to that the fact that all paranormals were women, and all the macho baggage *that* brings along...

Rob really was a gem. To him, it was just business. This time, though, business wasn't reason enough.

"Rob... I'm sorry."

"Look, Jen, I know you haven't had much to do with children—"

"Try nothing."

"Fine. Nothing. But, look – we have to assume the Piper knows this one's still alive. We need to move fast, before this bastard either gets away, or kills again to make up for losing one." He leaned towards me, his round face a study in the word *please*. "The truth is, Jen, the media's getting bored with the Piper. People just don't stay interested in street-kids. It won't be long before the department pulls the plug. Then it'll all just be so much more city noise, and I'll be pushed on to the next criminal flavour-of-the-day. We *need* this break."

I looked at the boy. Thrown on the streets by disease, or violence, or just indifference. A life wasted, before it had even begun. It could maybe be turned around, but the odds were definitely not on his side.

I puffed out a sigh. "There's a good chance I wouldn't be able to do much for you."

"All you have to do is try." Smelling victory, he added, "That's all any of us can do. But, hey, who knows? It might be what makes the difference."

I thought again about tonight's prescient episode, about my stomach, then looked at the boy again. One last sigh, then I threw gut wariness out the window. What the hell.

Besides, the cops paid well for murder consults, and the rent was coming due.

"Alright," I said. "But no guarantees, got it?"

The walrus smiled again. "Thanks, Jen."

* * *

Dana Harzberg looked up from her pocket nurse organiser, and said, "Jennifer, I wish you wouldn't do this."

As a Certified Attending Physician, Dana was as familiar as any non-paranormal could be with the methods – and dangers – of deep readings.

"I'm not any happier about this than you are, Dana." I lay down on the gurney that had been placed beside the boy's bed. "But it has to be done, so let's call your objections officially noted and carry on, okay?"

"Fine. But I'm telling you right now, children are vastly different from adult subjects. If I think things are getting out of hand, I'm bringing you back up." She gestured at a syringe of epinephrine.

"Just don't flinch, okay? Remember that I'll be at level four." I tried to look confident. "That deep a reading gives me a lot of control."

Dana curled her lip most eloquently, said, "Uh-huh," and plugged an IV tube into the port already stuck into my arm, starting a saline drip. Then she fiddled with the auto-nurse that would monitor my vitals, plugged the cables from it and the IV controller into her palm-top computer, and booted it up.

"When did your period end?" she asked.

Rob made an uncomfortable sound. "Look, if you two want some privacy—"

"Oh, Rob, sit down and shut up. I need you here as a witness." I turned back to Dana. "Two weeks ago, yesterday."

She entered that and some other things into the computer. The IV controller's little peristaltic pump began to whirr, feeding a mixture of hormones and RKT – a drug based on the beta-blocker propranolol – into the saline reservoir. Immediately, my hands and feet tingled and went cold. I made myself more comfortable, grabbed the boy's thin wrist, and waited for Dana's signal.

She frowned at the computer's display, then said in a voice like wind through a long, steel pipe, "Okay, Jennifer. Whenever you're ready."

I relaxed, centred... then, like a diver off of a high board, plunged down through the levels.

A swirling kaleidoscope of distorted, nightmare images slammed into me, flinging me about like a leaf in a gale. I fought for orientation, but the storm of twisted shapes and figures was overpowering – a Bosch nightmare, brought to frenetic life. I finally resorted to bulling my way through, tearing apart the encroaching images at their seams. All at once, they vanished. A wave of disorientation swept over me...

... then I was washed in bright sunlight. I blinked, glanced around. I stood on a manicured lawn, among deep pools of shade thrown by scattered trees. But it was utterly silent. Even the wind tossing the leaves made no sound.

I paused, catching my psychic breath, while a distant, analytical part of me recited: first layer, superficial echoes of trauma; second layer, protection through fundamental denial and withdrawal. Right out of the textbook. Somewhere, I should find a more explicit metaphor...

There. The boy sat on a swing, eyes cast skyward. I walked over, deliberately stepping between him and the sun.

No surprise. I cast no shadow; my hand wouldn't ruffle his hair. In this safe place his mind had created, I didn't exist. Just the grass, the trees, and the sun, all idealised to the point of plasticity. But no bruises, no pain, nothing that hurt.

I walked on, glancing back at the too-peaceful scene. Then I was back among the trees, and lost sight of him—

I was lying in a bed. Someone was holding my wrist. There were other people in the room. But it was blurred, unreal, just another fitful dream.

Next layer, basic motor-sensory functions. Off to the side, not quite accessible at this level, were the essential autonomous functions. It was I/him now, an assemblage of us both. I/he was beneath the upper protections, but this layer was just mechanics. There was nothing to learn here. I/he kept going, down...

I/he screamed.

There were monsters here, repeats of some of the things in the most superficial level. But now they were far more vivid here – and they were *aware*. They could see me/him, and they were hungry. They lurked under the bed, in the closet, just outside the bedroom door... and they were in the hall at the top of the dark stairs and down in the cellar and behind the shower curtain and they were EVERYWHERE—

I/he surged forward, through this uppermost layer of subconscious. The I-part was furiously erecting controls along the way, afraid that these terrors might sympathetically dredge up old Jennifer-childhood nightmares. Those, I/he may not be able to properly control here.

Deeper still. The monsters faded into a pervasive fog behind us. I/he hesitated.

The fog darkened, thick with trauma. In an adult mind, this was where the real dangers would begin. The he-part resisted going any further, afraid of exposure to something from which his mind sought, in its elemental wisdom, to protect him. But if the identity of whoever had done this was anywhere, it was in there. Give up now, and the killing would continue.

No more time. The I-part drove towards the darkness, dragging the he-part on, into the lower depths of the boy's subconscious mind.

I/he melted, melded, became we.

A man loomed over us, a black silhouette haloed by searing light. We tried to move, but he was kneeling on our chest. We could barely breathe, his weight crushed, pressed us down... we tried to push him off, to move, to breathe, but he raised his fist and slammed it down, BANG our head rang like dropped metal pans and the world flashed green. We tried to cry please don't please stop but we couldn't, and he didn't...

The nightmare memories went on and on, playing themselves out, while the part that was the emotional me separated a fraction and cried in sick guilt and helpless outrage.

But the flip-side, the distant, analytical me shouted in outrage of its own, appalled at this self-indulgence. This wasn't just a chance to wallow, commiserating in the boy's horrific memories; this was a chance to learn. I disengaged myself from boy a little further, enough to gain an external perspective, then wiped my mental tears and... watched.

Eventually, the pattern I sought resolved itself. The memory sequence was much more than a linear progression of images. Each portion of it carried along other memories, and those carried still others, in a branching network infinitely more complex than the most intricate of spider-webs. I forced myself to see past the horrible memory itself, and into the branches beyond.

There were so many... what was I looking for?

No... not what. *Where. Where* was this happening?

I pushed deeper into the web. It was... a room. And the room was connected, in turn, to a multitude of memory-images. But that one, there, of stairs; it led up, into a hallway... which led to a door... and into another room, with a stove. The stove had a clock, and a – no, ignore the stove, don't get distracted. It was a kitchen... and now I could see the path I wanted. A kitchen in a house, a large house, behind a hedge. A house number...? No, we hadn't seen any. Fine. We were in a car, with this man, driving down a street. He'd given us some money, and promised more if we'd... no, ignore that. Look for... there. A sign... with a street-name...

The pressure of denial swelled to an intolerable level. I had all I was going to get, without doing permanent harm. A man, in a house, on a street, with a name.

No, wait. Not quite all.

First, this man. We only saw his face in splintered glimpses, but that was enough. He was horrible, an evil cartoon-thing of bulbous eyes and thick, rubbery lips. Distorted by memory, no doubt, but I burned every line and blemish of it into the me-part's memory anyway, so that it would not, could not, be forgotten.

The other thing was harder to find. It finally turned up, a tiny bundle discarded in a far, dark corner. The boy's identity, his memories, his own, unique world-view...

... his name.

On the way back up, I paused before the swing in the silent park. Tears streamed down the boy's face, because something terrible had reached into this, his safe place, and hurt him so very badly. I wrapped my arms around him and whispered, "Oh, Kevin."

He stiffened and drew back. Our eyes met.

For a little while, anyway, he wasn't alone.

* * *

"Jennifer, what colour is this light?"

"Blue. Now it's red."

"Good. How do you feel?"

"Uh... fine. Usual light-headed feeling..." My tongue felt twice its normal size.

"How is she?" Rob asked from somewhere off to one side.

"I *think* she's alright," Dana grouched, without looking up from the pocket nurse. She entered something in her computer, then turned it off. "Jennifer, what happened in there?"

"Oh, Dana, it's so bad—" I began, but Rob stepped forward.

"Jen, did you find out anything we can use?"

I turned to him and opened my mouth to say, yes, I know where this son of a bitch is—

—but bit off my own words.

The police weren't going to catch him.

It wasn't a feeling, or an opinion; it was a fact. I didn't know why they'd fail to get him. There was no context, no explanation – just an isolated fragment of certainty, like a snapshot. But there was no doubt. I just knew it, in the same way that I knew I was lying on a gurney in Sick Kid's hospital in Toronto.

Prescience.

Again.

I fumbled for something to say. "Uh... I..."

Dana frowned. "I don't think she's completely recovered yet. Let's give her some more time."

That was fine by me. I needed time to think. Rob closed his mouth and nodded.

"Okay. I could use a coffee, anyway."

After they'd gone, I looked at Kevin. The bastard who had done this was going to get away with it.

Shit! It wasn't right...!

No... wait.

A thought drifted by, just out of reach. I screwed my eyes shut and tried to ignore the dentist's drill whining behind my eyes. What was I thinking...?

What I *knew* was that the police weren't going to catch him. But that didn't mean he was going to get away with it.

A man, in a house, on a street with a name.

I knew what I was going to do. And this time, prescience had nothing to do with it.

* * *

I peered through the darkness, dithering. This had to be the right house. A few of the others were close, but none fit as well with the one in Kevin's memory. A red sandstone Victorian, that said *old money*. I glanced around, taking in the graffiti and wind-stirred trash. Well, maybe this was old money, once. Now it was just old.

I checked the elastic band holding the pneumatic syringe on my right forearm. The sequential doses of RKT and epinephrine that filled its cylinders weren't tough to get, if you knew the right people. Neither had any real street value. But RKT was meant to be used clinically, for focusing and boosting paranormal abilities – not on the fly, without doctors like Dana watching over things. I'd meant it as a sort of insurance, or maybe a psychic placebo. In reality it was more likely to kill me, than to be of much help.

But it didn't matter. I wouldn't need it. There'd be nothing to this. I stepped out from the shadows under the hoary old elm, took a deep breath, and walked up the cracked flagstones.

I thought again about calling the cops, and again decided no. Prescience might sometimes be short on the 'why' or 'how', but the 'what' was always accurate. If I called the cops without being sure of the house, they'd start poking around the neighbourhood, and that might be the very reason they wouldn't get him. No. This was better. If I could corroborate what I'd learned in Kevin's memory with a man in this house, then that might change my prescient mind – I hoped. And if not... well, then I'd tell Rob what I knew, and just hope for the best.

Front door. A siren wailed from off towards the Toronto skyline, and I glanced that way. A misty childhood memory drifted by... that same city skyline, brightly lit against the night sky. Huh. Not anymore. These days, except for a few strobing anti-collision lights lit for the benefit of airplanes, the buildings were just dark shapes, like blackened teeth sprouting from smog-hazy, twilight gums.

The siren stopped, and silence pooled back around me. Okay, just knock, see who answers. If it's him, make an excuse, leave. That's all.

I raised my hand, knuckled it. Hesitated, then tapped it against the door.

Nothing. Then a thump, and a hall light clicked on. A pause. A head passed across the glowing rectangle of window. Another pause. My stomach did a slow roll.

The door opened.

I was facing a woman, thirtyish and tall, with a brunette page-boy framing a square, almost masculine face.

A woman...?

She peered over the door-chain. "Yes?"

Uh... My mind raced. Finally: "Is your husband home?"

Oh, what a stupid thing to—

"My husband?" She frowned. "Who can I say is calling?"

"My name's Jennifer. Jennifer Platt." My mind whip-sawed from side to side. "I... I found a wallet today... and it might be his, it had this address in it, and... well, I'd like to show it to him."

Christ, that was so pathetic.

She beamed a long, searching look at me. I focused hard on looking benign and casual, and forcing myself to not fidget. Finally, she shrugged.

"Come on in. I'll go get him." She unlatched the chain and pulled the door open.

The hallway was lit by a single fixture. There were stairs up, a closed door at the far end, and an open one to the left that led into a living room flickering blue with TV light. I stepped onto the mat, but no farther.

"Thank you," I said, regaining some composure. "I'm sorry to trouble you, but I was on my way home from work, and thought—"

"That's alright," she said, smiling and offering her hand. It took it, felt cool flesh, released it. "I'm Mary," she went on. "My husband's upstairs. I'll be right back." She padded up the stairs, her socked feet silent in the mangy tan shag. "Honey, someone to see you...!"

I took a few deep breaths while she was gone, clearing my head. Okay. When he comes down, take a *good* look. Reach for the wallet, then say, oops, left it in my car. Be right back. Out the door and gone. No problem.

Another breath, and I tasted the ghosts of old dinners. No problem?

Yeah, right. Tell my still-rolling stomach that. After this, stick to the paranormal work, and no more of this detective nonsense.

I smiled. Jennifer Platt, detective. Yeah. Well, Ms Platt, if you're a hotshot detective, what can you learn from the scene? Hmm. Okay. Shoes by the door. Sneakers, pumps, and sandals. So, Watson, the perpetrator obviously is none other than…

Huh. Sneakers, pumps, sandals. The pumps were peach-coloured, a little scuffed. The sandals were definitely a woman's. The sneakers, well-used and grubby, were small… men's sixes, no more… again, probably a woman's.

Maybe I should try reading her. Risky, but I could probably do a brief level one without attracting any—

Oh. Shit.

She'd been as quiet coming down the stairs as she had been going up. Except now she had a gun.

"Well, Jennifer Platt," she said, "I'm afraid my *husband* can't come to the door right now."

"I—"

She shook her head. "Don't. Don't say anything. Just go. That way." She stepped down another step and jerked the gun towards the living room.

I took a step, stopped.

"Look—"

"NOW!"

I backed into the living room and found myself surrounded by tatty, mismatched furniture. A dark arched doorway framed a cheap metal dining room set, all of it lit TV-blue. The woman named Mary followed me, the gun waving with her steps. I thought about my phone, and then about how I'd be dead long before I could even begin dialling it.

"Look," I said, "I just came to—"

"Oh, the wallet. I know." She raised the gun until the sight bisected her eye. "You can drop the bullshit. You came here to spy. For the police."

I swallowed, shook my head. "Listen." I swallowed again. "They're —"

"— not anywhere nearby. Pretty stupid coming here all alone."

How the hell did she know that—?

Oh, shit *again*. She was a paranormal.

And I'd shaken her hand. Like I'd thought only seconds later, level one was barely noticeable.

She gestured with the gun again. "Sit down."

I glanced around. A chair beside me. I sat.

"You read that boy, the one that got away," she said, nodding. "I knew he was going to be trouble. I knew it, I TOLD YOU!" I jumped, although her last words, nearly shouted, didn't seem to be directed at me.

She shook her head. "Doesn't matter." She frowned. "So why you? Why not the police?"

I opened my mouth. But what could I say?

"You're hiding something," she went on. "You're hiding something, I read it, and I WANT to know what it IS!"

I was hiding lots of things, but I guessed she was talking about the syringe. It'd been in the back of my mind, but probably too deep to read at level one.

Abruptly, she shrugged. "It doesn't matter. I know you came expecting to find a man. And that's good."

"Good...?"

"Uh-huh. See, I live here alone."

"Alone? But..."

And that was the answer.

The man in Kevin's memory didn't exist. He'd been planted there by this woman, overprinting her own image. That explained the cartoon-evil face. It was a caricature, something she'd manufactured. It told me again how strong she was. As I understood it, memory alteration was, at best, a transient thing. But she'd made it stick.

It also explained why the police wouldn't catch him.

There was no *him* to catch.

I looked at her, trying to ignore the blackness of the muzzle. She gazed back, her dark eyes hard as flint.

"It's you," I said. "You're the Pied Piper. You're the one who's been killing those children—"

"NO!"

I jumped again.

"I've never hurt anyone!" She leaned forward. "I tried to help those poor children. Really, I did."

I didn't believe her, of course, but nodded anyway. "Okay. Do you know who has been killing them, then?"

"My father."

"Your father."

She nodded. "That's right. He always hurts children. Always."

"I see." I glanced at the gun. "Where's your father now?"

She lifted a finger, tapped her head.

"Right here."

I stared.

"You see… my father… he hurts children. It's what he does. He hurts them because it makes him feel so strong, so… so powerful. And he's still doing it, even though he… he's dead." Her voice collapsed into breathy sobs. "Even though he's dead, he… he's *still* hurting them."

Psychotic. Completely psychotic. Combine that with her paranormality and I couldn't have invented a more dangerous predicament. I opened my mouth but couldn't think of anything to say.

"Do you know what the worst part is? He makes me help him. I only want to help these poor children. They're getting hurt and dying out there on the streets, from drugs and disease and accidents. From all kinds of things. And nobody gives a shit." She smiled suddenly, brightly. "Except me. I care."

I nodded. "I'm sure you do. But—"

"I DO! I KNOW YOU DON'T BELIEVE ME BUT I DO!" She was half-standing now, both hands clamped around the gun. She took two ragged breaths, three, and bile surged acid-acrid in my throat. In an instant, the gun would explode in my face…

But she sat back down and resumed the sweet-sweet smile. "I really do care. My father, though…" Her face darkened and she shook her head. "He makes me… makes me do things to them. He makes me hurt them, and I can't stop him… I've never been able to STOP him."

I took a shuddering breath. My life depended on what I said here. Try something innocuous…

"Look… Mary, maybe I can help you."

"Help me…?"

I forced my eyes up from the gun, to hers. "I can help you stop your father, Mary. Stop him from hurting any more children."

A puzzled frown creased her face and the gun wavered. "How could you stop him?"

"By taking you away from him, Mary." I wanted to keep repeating her name. "So he can't hurt you, or make you hurt anyone else."

The gun wavered a bit more. "You can do that? Make him go away?"

I nodded. "With some help. But first you have to give me the gun, Mary."

"Give you… Oh, no. No, I don't think so."

"It's alright, Mary. Give me the gun, and then we'll get you some help." I put out my hand. "Give it to me, and we'll get you away from your father."

Her eyes flicked over to my hand, and her face softened. My hand was a way out for her, all she had to do was take it—

14

Something on my upraised arm glinted in the blue TV light.

The syringe.

Mary's eyes went crystal-hard.

"YOU LYING BITCH!" She jumped to her feet and jammed the muzzle of the gun into my face. "You didn't come here to help me, you came to DRUG me and to KILL ME!"

"Mary, NO—"

Her other hand snapped up in front of me.

"Give it to me," she said.

"Mary, please—"

"GIVE IT TO ME!"

My eyes never left her finger, bloodless-white around the trigger. I pushed up the sleeve of my jacket, felt for the syringe, pulled it out from under the elastic. For one wild second, I thought of plunging it into her hand and just hoping for the best. But the first injection, the RKT, wouldn't have any effect before she could pull the trigger and blast my head to bloody bits. I finally just handed it to her, and she snatched it away.

"It's time to decide what we're going to do with you, Jennifer Platt. Me, well, I'd like to just let you go. But my father..." She shook her head. "I'm sorry."

Her finger tightened. In another second, my world was going to end.

I didn't plan it... didn't even think.

Just drove my hand up and snapped my head to the side. Mary reacted reflexively, yanking the gun back and—

There was heat and a flash and a loud metallic CLICK, and something whumped past my ear. But I wasn't after the gun. My open palm connected with the syringe instead, driving the needle back against its stop. The pneumatic cartridge hissed, pumping the RKT into my hand. I kept going, sideways now, crashing against the table beside me and sprawling onto the floor.

Mary screamed something and fired again. The shag beside my arm puffed up a cloud of carpet-fibres and dust. I dove, crawled, heading for the gloom in the dining room, expecting a hammer blow in my back, pain exploding, breath blown out of me, then darkness swirling up as the bullet tore through me...

Click-WHUMP, splinters flew from the door-frame as I stumbled through, click-WHUMP into the floor, Oh God I couldn't move fast enough, she was right behind me click-THUMP into the carpet click-CLANG-WHIRR off a metal dining-room chair that's it dead end I spun she was silhouetted against the

TV glow just feet away the gun centred on me as I lifted a hand gone ice-cold numb lifted it so slooow...

... brushed her leg...

... linked.

She was strong. But now, I was stronger. I rode the crest of the RKT wave deep into her mind, level three, four... five...

Motor-sensory, the blade-sight planted in the middle of Jennifer Platt's face, finger squeezing the trigger, hammer rising, falling—

We jerked the gun aside flash-click-THUMP, the muzzle-blast swept Jennifer Platt's hair, the bullet smacking into the wall behind her.

We lowered the gun.

Father screamed, his face no longer cartoon-evil, just drink-flushed and rage-contorted, the way we'd seen it so many times, late at night, crashing into our room, kneeling on our chest, we could barely breathe, his weight crushing, pressing us down, and we tried to push him off, to move, to breath, but he raised his fist and brought it down, BANG our head rang like dropped metal pans and the world flashed green. We tried to cry, please, don't... please *stop*... but we couldn't...

Eventually, we ran, lived on the streets, sold our body under the cold neon, until the police found us, took us, brought us home.

Then did it all over again.

Then the day came when we watched them bury him, then came back to the house that was ours now, and here he was, waiting. Children, ugly, ugly little children had to suffer, he said, because that's what children did. They suffered.

Back onto the streets, he took us to the kids who had no one to trust, to the kids who sold their bodies under the cold neon.

They suffered, they all suffered...

No one to trust.

No one...

Until now.

Mary could trust Jennifer Platt.

Absolutely.

Trust her to withdraw, to leave her alone in her mind.

Trust her, to let her put the gun into her mouth.

Trust her, to let her pull the trigger.

* * *

"Stupid stupid STUPID!"

"Say it again, Rob."

"Stupid!"

I nodded. The movement still hurt, but then, I was only a day out of the RKT-induced shock.

Not meant to be used that way, without buffers and stabilisers, Dana had lectured. Practically poison. I winced again. She was right.

Rob leaned on the bed-rail. "Promise me, Jen, that next time you let me do the cop shit."

"I promise."

He sighed, long and slow. "Well, as it turns out, the physical evidence from the house definitely ties her to the boy. I'm not sure it'll link her as well to any of the other victims, but I don't think it's going to matter. Barring any recurrences, this damned Pied Piper case is finally closed."

"I'm glad."

"So am I." He shook his head. "Christ, a woman. I never would have thought a woman..." He puffed out a breath. "She must have been pretty badly screwed up."

"She was."

"Huh. Well, lucky for you. If she hadn't flipped out totally and decided to blow her brains out when she did—"

"Rob, I'd... I'd like to rest. Okay?"

He sighed again. "Yeah, sure. I'll see you later."

"Later. Okay."

"Rob?"

"Huh?"

"Is Kevin still in the hospital?"

"Yeah, I think so. Community and Social Services is looking for a foster placement for him. Why?"

"Just wondering. Thanks."

He opened his mouth, then shook his head, closed it, and left.

I sank back onto the pillow.

If she hadn't blown her brains out...

I could have stopped her. Maybe I *should* have stopped her.

But I remembered the last thing I'd felt before I'd left her mind.

Gratitude.

There was finally someone she could trust, someone who would let her have control over her own life, even if it were only for one, final instant.

I thought about Kevin. Fostered and forgotten, he'd just end up on the

streets again.

Suffering under the cold neon…

No. That was something I could try to stop.

I queried the directory on my phone, got it to search out the number for Community and Social Services. As it worked, I thought about Mary… and again, about Kevin.

What was it Rob had said?

"… *what we all have to do is try*…"

I looked at the phone number.

I still didn't know kids.

But that could change.

I pressed 'DIAL'.

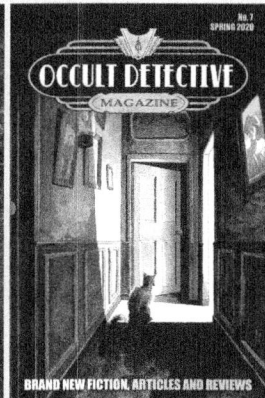

THE BONES ARE WALKING

REBECCA BUCHANAN

The streets of Port-au-Prince were a wild jumble of colors, scents, and sounds. Banners and flags hung from every lamp post, window, and railing; they were even draped over the backs of horses and the hoods of rattling automobiles. Vendors hollered, offering bottles of liquor, dried tobacco, peppers, and sweets. Women with babies strapped to their backs and men dressed in cool, bright cotton wove their way among the stalls, haggling over food, flowers, and anything else they thought would make their celebration perfect and the envy of their neighbors.

The carriage rocked as a brightly-dressed houngan marched past, leading his congregation; many of the devotees carried platters and baskets of cassava, peanuts, maize, and eggs, as well as bottles of rum and fat rolls of cigars. Other houngans and mambos led their own companies of celebrants through the stone-paved streets, back and forth, in every direction, moving between the Palais du Gouvernement, and the numerous hounfo and cemeteries scattered throughout the city and beyond. Individual Garde milled around, pistols and batons at their hips as they watched for thieves and pickpockets hoping to take advantage of the festivities.

I should be celebrating right now.

The sour thought took Anaïs by surprise. Certainly, she had been looking forward to joining her mother, aunts, uncles, and so many cousins that she couldn't keep them all straight at the ancestral hounfo in Kenscoff.

One hundred years since the French had been driven from the island's shores and La République had been established. A century of hard-won, hard-preserved freedom and prosperity, despite numerous attempts by European and American powers to take the island.

The twentieth century had arrived only a few years ago in an explosion of fireworks, sacrifices, and week-long revels. No one who had seen it would ever forget.

And yet independence celebrations were already looking to be even more spectacular.

The carriage rocked again as they turned a corner, slowing to allow the crowd to thin. The driver snapped something and urged the carriage forward. Anaïs could hear dogs barking madly and children giggling and

shrieking. A bloom of spicy and sweet scents – riz collé aux pois, tchaka, pikliz, mangoes, bananas – rolled through the window and set her mouth to watering and her stomach to grumbling.

There would be no celebrations for her. Not yet, anyway. She was here, performing her oath-bound duty, as was right and proper.

"So what are we going to do about the ghost?" she asked.

Across from her, Judge Celestin remained silent, his gaze fixed on nothing and everything outside. His black silk top hat, with its wide purple ribbon, sat straight on his head. His black tuxedo was neatly pressed, the white lace at his throat and wrists spotless. Dark spectacles hung from a purple ribbon at his lapel. His purple leather gloves were brand new and creaked slightly as he tapped a finger on the head of his walking stick – the only outward sign of agitation.

Without turning his gaze from the window, he asked, "What do you think we should do, mon apprentie?"

A test. Of course he would turn this into a test.

Anaïs suppressed a sigh and tugged at one sleeve. Her own suit was solid black. She had not earned her purple yet, or even her top hat, and only Baron Samedi himself could say when she had done so.

The Judge turned towards her. "Assuming, of course, that it is a 'ghost,' as you say."

She bit the inside of her lip.

There were other reasons to summon a Judge, though they were far less common. In the eight years since Baron Samedi had possessed her at a Fèt Gede and announced that she would apprentice to his most famous servant, none other than Judge Alexandre Christophe Celestin, she had encountered only a handful of these other reasons – and she had barely survived two of them.

Celestin snorted, his eyebrows drawn together in a dismissive frown. "Ghost. Feh!" he snorted. "You have been associating with those Irish parese moun too much. Use the proper term."

She dipped her head. "Gede," she corrected. "I apologize, mesye. And those Irish students are not lazy. They are curious and want to know about us and our land and our ways."

The Judge rolled his eyes. "So they spend their days sitting in cafés drinking coffee and rum, while they ogle the pretty boys and girls who walk past. Feh!"

The carriage rocked again as it turned another corner. Anaïs peered out the left window, catching sight of the Palais du Gouvernement through the

crowded street, the wrought-iron fence, and the thick stands of ornamental mahogany and cedar trees.

The building had been much changed and expanded in the century since it had been the seat of the French colonial governor. As the carriage pulled past a company of Garde and through the gates, Anaïs could see the wide mahogany stairs leading up to the soaring white columns of the residence of Présidente Chevry. The portico had been draped in flags, with patriotic pots of flowers and temporary shrines to the Lwa scattered along its length. To either side, the wings housing the Parlement Haïtien stood empty—

Anaïs squinted and leaned further out the window, peering around the carriage and towards the mass of people gathered near the steps of the Chamber of Deputies to the left of the presidential residence. A dozen Garde stood in twin ranks in front of the door, with more scattered around the perimeter of the building. A small mob in the suits and fine dresses of high-ranking officials, not the festive clothes of celebrants headed off to an independence party, huddled in the circular driveway. Half-a-dozen military automobiles and as many horses stood in the shade of the trees nearby.

Celestin stuck his head out of the window in front of her, blocking her view. One hand held his top hat in place, the lace at his wrists dancing in the wind.

Lips tight, he turned to look back at her. "Not a gede," he said, and disappeared back inside the carriage.

* * *

"There is a gede," Monsieur Travieso announced.

He had identified himself as Présidente Chevry's proxy; an advisor of some sort who stepped in when the Présidente was too busy with other matters to deal with... well, whatever this was.

"Make it leave," Travieso continued. He glared down at Judge Celestin, his arms across his wide chest, his chin thrust out.

Behind him, the mob of officials shifted uncomfortably. Many exchanged worried glances or wrung their hands or fiddled with buttons. Only the Garde Kapitènn appeared unmoved; her expression was grim, her hands clasped behind her back, a pistol on either hip.

Anaïs gaped at Travieso. "*Make it leave??*" she stuttered.

He didn't even glance in her direction, his eyelids twitching as he continued to address the Judge. "It is causing a scene. It is embarrassing. Ambassadors from the world over – Britain, France, America, Russia, as far away as Japan, even — will be here soon to witness the centennial of our independence. We cannot afford to have this... this... *nuisance* humiliate us in the eyes of the world."

Anaïs continued to gape.

The Judge tilted his head slightly. "They do enjoy a good party."

Travieso's eyes narrowed.

"All of the best parties have at least one gede; though rarely is it only one. Why I once attended a wedding where the deceased great-grandparents and great-great grandparents and great-great-great grandparents all insisted on making an appearance. Their reminiscences about their own wedding nights — and their advice on how best to enjoy the marital bed — were quite entertaining."

Travieso's jaw was so tight that a muscle spasmed.

"But, as I said, not a gede." Celestin twirled his cane, revealing the phallus-shaped head, and set out past the ranks of Garde and towards the great double mahogany doors that led into the Chamber. "Come along, mon apprentie!"

Giving Monsieur Travieso one last disbelieving look, Anaïs hurried after the Judge. Feeling a presence, she looked over to see the Garde Kapitènn walking along at her side, strides long and determined.

"Kapitènn Massez." The woman dipped her head towards Anaïs, the grey in her tightly coiled hair visible now that she was closer. "Under the circumstances, I think it best that I accompany you."

Anaïs studied the Kapitènn, noting the tight lines around her eyes and the blood splatter no longer hidden by the dark green of her uniform. "... Very well."

They quickly reached the Judge, who had ascended the short flight of stairs and stopped, one hand on the wrought iron handle of the door. He half-turned towards Anaïs and Kapitènn Massez. "He is French, wi?"

"Travieso?" The Kapitènn nodded. "Wi. Paris, born and raised. He immigrated to Haiti about ten years ago."

"And he, a non-native son of La République, became such a close advisor to Présidente Chevry, how?"

The captain shrugged. "They worked at the same hospital. When Chevry left her practice and entered politics, he followed along." Another shrug. "He's arrogant, small-minded, and quite often an ass. I truly think Chevry keeps him around out of sentimentality. Perhaps she knows that he would not survive long on his own."

Celestin *hhmmed* at that. Turning away, he started to push open the large mahogany door, but the Kapitènn stopped him with a word.

"Judge Celestin." She hesitated, continuing only when he turned back towards her. "Monsieur Travieso does not understand Haiti or its spirits. Don't startle her. I've already lost a dozen Garde."

Anaïs frowned. *Startle her?*

Celestin did not answer. He pushed on the heavy door. It creaked slightly as it opened.

Beyond, the Chamber of Deputies was unexpectedly dark. Thick curtains had been drawn across the floor-to-ceiling windows. Only a few electric wall sconces cast feeble light over the tiered seats that lined each side of the room and the podium at the front.

But the sunlight that spilled in through the door and across the mahogany floor was more than enough to see the blood and the mangled bodies of the Garde, and the injured Présidente Chevry, and the once-human-thing that loomed over her, and the monsters that prowled back and forth between the shadows.

Anaïs held her breath, silently reciting a prayer to Bondye, Baron Samedi, Brav Gede, and Manman Brijit.

The Judge took a step into the room. Kapitènn Massez followed, a pistol in each hand, and then Anaïs. She unbuttoned her jacket, allowing the clothing to hang open so that she could quickly reach the ritual items stored in the inner pockets.

She noted, with silent alarm, that Celestin had not done the same. Instead, he continued forward, his cane clack-clack-clacking against the wooden floor. Only when he was a dozen steps from the Présidente and the once-human-thing that stood over her did he stop.

The door closed behind them with a muffled thud.

"Judge Celestin." Présidente Chevry dipped her head – as much as she could move from where she sat on the floor, one knee twisted at a wrong angle. Her face was bruised and bloodied, her brilliant white dress torn and splattered with red, her ceremonial sash in tatters across her chest. Still, Chevry held herself straight, and it was not fear that lit her eyes, but anger. "I apologize for summoning you away from the celebrations."

The Judge removed his top hat and swept it across his chest as he bowed. "No apology is necessary, mon Présidente." As he rose, he carefully set his dark spectacles on the bridge of his nose.

Anaïs had worn his spectacles. Once. He had been injured and there had been no choice.

The world as seen through those glasses... all garish colors and wandering spirits and things that had never been human and things that were no longer human. The world as it truly was.

She shivered, pushing the thought away, focusing on the Work at hand.

The thing that loomed over Chevry hissed — a high-pitched near-shriek that set Anaïs' teeth on edge. Her hands twitched, longing to fling handfuls of fresh basil and spring water and a blessed dagger or two at the thing.

It might have been female, once, if the sagging mounds on its chest could be considered breasts. Formerly dark skin had turned splotchy with decay. Its cheeks and eyes were sunken, and long straggly hair hung down its back.

"Bring him to me," the thing hissed.

The monsters circled them. Anaïs watched them from the corners of her eyes. Just as the thing had once been human, so they had once been animals: boars, snakes, bats, crows, rats, even caimans. They were baka now, malevolent spirits given a home in the body of an unwilling sacrifice. There were dozens of them, snapping and chittering as they crawled across the floor and over the walls and peered down from the darkened chandeliers.

As they drew closer, Anaïs spun on her toes and carefully backed towards Celestin. Kapitènn Massez did the same, her pistols raised. Anaïs slipped a hand inside her jacket, pulling out a pair of thin daggers that hid neatly within her palm.

"Bring him to me!"

"Who?" Celestin asked.

"My child. My sweet boy." The thing crooned, a grotesque sound coming from its emaciated chest. The sound turned harsh. "It took him. The jé-rouge took him!"

Jé-rouge.

Anaïs swallowed hard. Beside her, Massez muttered an oath under her breath.

Hearing a gasp, Anaïs risked a quick peek. The thing had wrapped a long-fingered hand around Présidente Chevry's throat. A caiman lurched forward and Anaïs snapped her focus back to the baka that surrounded them. She slapped her foot hard against the floor, driving it back a few steps.

"A jé-rouge?" Celestin said slowly.

Another gasp from the Présidente as the thing snarled. "Sunset. Bring him to me by sunset."

Celestin made a low grunt. "The jé-rouge will not even reveal itself until dark. I cannot find it before then."

hhisssssss

Présidente Chevry gurgled.

The caiman lunged for Anaïs' foot again. She twisted to the side and flung one of the little daggers straight into its right eyeball.

It twisted and rolled, Manman Brijit's blessing eating away at both the malevolent spirit and the corrupted flesh that it inhabited. The baka shuddered, whined, and stilled. The body of the caiman turned grey, turned to powder, slumped, and crumbled across the floor.

Up in the chandeliers, the crows screamed.

Celestin slammed the point of his cane against the floor. The crack echoed, echoed, echoed from the walls, making Anaïs flinch.

Silence.

"Midnight," the Judge said, his tone flat and uncompromising.

hhiisss "Midnight. For every hour you are late, I feed a piece of her to my pets."

From the corner of her eye, Anaïs saw Celestin tip his head in agreement.

"You have something of the child's?" he asked.

The once-human-thing snarled and twitched her – its – hand. A pair of rats scampered forward, carrying something in their jaws. Anaïs could not see it clearly. Whatever it was, the Judge stared down at it for a long moment; then he pulled a white handkerchief from a pocket and bent to pick it up.

Celestin bowed. "Mon Présidente."

"Judge Celestin," Chevry whispered, her voice scratching. "I trust you to hold to your oaths and your duty."

He held still for a moment, silent. Then Anaïs felt him back towards her. She began moving slowly towards the door, Massez matching her steps.

The baka followed, growling and chittering from the shadows.

When they were close enough, Anaïs reached out and pulled open the heavy mahogany door. And then they were outside, blinking in the bright sunshine and heat of midday.

The door thudded shut behind them.

Anaïs drew a deep breath, filling her lungs with the scent of trees and sea and city.

Travieso appeared suddenly, his jaw still thrust forward. His angry gaze was fixed on Celestin. "Well? Did you get rid of it?"

The Judge ignored him, white handkerchief still in his hands though he had wrapped it around whatever the once-human-thing had given him. "Kapitènn, we will return by midnight. Until then, no one is to enter the

Chamber of Deputies."

Massez nodded, sliding her pistols back into their holsters. "Wi. Understood."

"Celestin! Are you heeding me?" Travieso demanded.

"Non. I am ignoring you. Come along, mon apprentie."

Travieso grabbed his arm.

Anaïs leapt forward, little dagger in her hand. She kicked the side of Travieso's knee, knocking him to the ground. He fell with a pained grunt. When she pressed a dagger to the soft flesh beneath his eye, he froze.

There were squeaks of alarm and surprise from the officials. Several of the Garde shifted on their feet, looking uncertain, until Massez waved them down.

Very slowly, Travieso released his grip on Celestin. Even more slowly, he leaned away from Anaïs. When she did not follow, her knife held still and steady, he grimaced and pushed himself back to his feet. This time, his gaze was on her as he smoothed his jacket and tried to hold himself upright by putting all of his weight on his good leg.

"You heard the Judge, Kapitènn," Anaïs said, slipping her dagger back into her jacket.

"Wi," Massez answered.

Without another word, Celestin and Anaïs walked down the steps, through the ranks of Garde and the mob of officials, and climbed back into the carriage. The Judge signaled the driver, and they rolled away.

* * *

"The umbilical cord." Celestin held open the handkerchief. It lay shriveled and dried in the center of the white cloth.

Anaïs leaned closer, across the space that separated their seats. She wrinkled her nose. "How could that *thing* have produced a child?" she asked.

"Yes, how?" he responded.

Another test.

Anaïs paused. "She was human. A mambo, perhaps, who turned away from the Lwa. Or a bokor who forged her own path. Could the child have been the reason?"

The carriage bounced, slowing as it passed through a crowd.

"Possibly," Celestin answered slowly, focus fixed on the umbilical cord. "Perhaps she wished for a child, and when one would not come to her through the blessings of Bondye, she sought the aid of more sinister powers.

Or the child was ill. Or stolen, never even hers to begin with. And now he is gone, taken away by a jé-rouge…"

Anaïs shuddered, leaning back in her seat.

What had that cute Irish student called the jé-rouge? Wolf something-or-other. But the creature she had described to Anaïs over a perfect bottle of rum had sounded very little like the Haitian monsters. The Irish wolf-creatures lacked red eyes, for one, and were benevolent protectors of women and children.

The jé-rouge, on the other hand, were malevolent entities who possessed their unwitting victims at night, and stole little ones away from their mothers while they slept.

"If I ask how the umbilical cord is useful to us, you will just turn the question back on me, won't you?"

Celestin grinned at her over the top of his glasses. "My apprentice is a clever girl. I am sure that you will be able to work it out."

Anaïs frowned at the gross thing in the handkerchief. She pressed a hand to the seat, holding herself upright as the carriage took a sharp turn. She glanced quickly out the window, taking in the familiar buildings. "We are going to your sister's hounfo. A servant's first step should always be to consult with the Lwa. And there is little that escapes the notice of the dead."

The Judge nodded. "Yes. Though I expect there will be more to it than just that." His expression turned grim as he folded the handkerchief and tucked it inside his suit, then dropped his spectacles on their ribbon. "There is a jé-rouge involved. It must be killed. And also the one who has fallen away from the Lwa, onto a path of misery and pain. Finding the child will only be part of the task they set us – likely the reward for taking care of those *other* matters."

The carriage creaked to a halt. The driver stepped down and pulled open the door.

Anaïs looked out to find they were at the entrance of the hounfo. Mambo Johanne stood on the front steps, her hands braced on her hips. Her long white dress drifted lightly in the breeze, and not a single strand of graying hair poked out from beneath her white head scarf. Her face was screwed up in an expression of affectionate exasperation and bemusement.

The Mambo shook her head as they stepped out. "I was expecting you. I could hear the whispers." She waved a hand vaguely, then squinted down at them. "You have a taint about you. Baka? Feh!" She turned away and strode up the stairs. "Come along, mon frère, mon amie. The Lwa grow impatient and we should not keep them waiting."

Anaïs dutifully trailed the Judge through the wide open door and into the cool and relative quiet of the hounfo. They passed through the open-air peristil, the large cedar tree that grew from the center of the courtyard providing shade and filling the space with its sweet scent; two brightly colored banners for the serpentine Lwa Danbala-Wedo and Ayida-Wedo spiraled up the trunk. The ground had long ago been pounded flat and smooth by hundreds of feet circling round the tree.

Half-a-dozen doors opened off the peristil, low voices and laughter and the rap of drums coming from some of them. Mambo Johanne led them to one on the left, where they removed their shoes and socks, and entered the altar room of the Gede Lwa. The black walls, floor, and ceiling were covered in rugs, pillows, banners, and streamers of various shades of purple, while brilliant white veves shone from the walls; the largest and most intricate sigils, those for Baron Samedi and Manman Brijit, were painted above their individual altars, the layers of tables and shelves covered in offerings of rum, piman, bread, black coffee, cigarettes, bonnets and top hats, dried flowers, little gilt mirrors, sequined bottles, and clay phalluses.

To the uninitiated – like that ass Travieso – it would have appeared a cluttered mess. But not to Anaïs and everyone else who served the Lwa.

She inhaled and exhaled slowly, feeling a tightness she had not even recognized slip from her shoulders and chest.

Home.

Now the real Work could begin.

The Judge and his sister shared a brief, whispered exchange. With an abrupt nod, Mambo Johanne exited the room. She returned moments later, leading a small company of drummers and hounsi. Barefoot, silent, heads bowed, they all sat on the hard ground, facing the altars of the Baron and his wife.

And then Mambo Johanne made the Sign of the Cross and began the Priyè Ginen. The beautiful, musical French of the Lord's Prayer, the Hail Mary, and the Apostle's Creed wove around Anaïs. Back and forth, round and round, calling out and answering back. Songs to Bondye and to the Blessed Virgin and to Christ and to all the Saints followed, the song swelling, filling the room and flowing out into the peristil.

More initiates joined them, attracted by the sounds of celebration. The room became hot and crowded.

The last of the beautiful French faded into the afternoon. Mambo Johanne picked up her rattle, the drummers taking up the beat. The words were in Kreyòl now, and langaj, faster, faster, the tempo accelerating.

Zo-a li mache, they sang.

The bones are walking.

Drumming, chanting. Mambo Johanne shook her rattle, inviting Baron Samedi and Manman Brijit and Brav Gede and so many others, all of the forgotten dead, the abandoned dead, the lost, to join them.

The celebrants all stood as one, still clapping, stamping their feet, snaking round to exit the altar room. They proceeded out into the peristil. The crescent moon shown down from above, most of the stars lost beyond the branches of the tree, the lights of the city, the flashes and smoke of fireworks. They circled round and round the cedar, clapping, calling, stamping their feet, inviting the Gede Lwa to join them.

Anaïs was close behind Celestin, her hair and clothes slicked to her body by sweat, when she saw him jerk. His whole body twisted, his back contorting to a seemingly impossible degree. And then he snapped back upright, top hat still on his head, and giggled.

It was the sweet, innocent laugh of a little girl.

Anaïs pushed through the hounsi around her, reaching the Judge at the same time as Mambo Johanne. Several of the initiates formed a loose circle around them, keeping the other celebrants at a distance for the time being.

Celestin-who-was-no-longer-Celestin plopped onto her bottom on the hard floor and giggled again. "I'm thirsty," she announced and waved her hands. "Piman! Piman!"

A bottle of the heavily-peppered sugarcane rum was handed to Anaïs, who passed it to the little girl. She squatted on her feet while the celebrants continued to circle around them, watching the little girl who had taken the Judge's body guzzle the entire bottle. Even with her head tipped all the way back, the top hat stayed in place.

The little girl smacked Celestin's lips and then grinned up at Anaïs. "I'm Lucille. Baron's busy," she said. "So's Manman. They told me to help, but they have Work for you."

"It is a pleasure to meet you, Lucille. I am Anaïs—"

"I know."

"What Work do the Baron and Manman require of us?"

The little girl wrinkled Celestin's nose in thought and rolled the empty bottle in her hands. Then she nodded. "Kill the jé-rouge. Kill the fallen mambo. Find the thief. Return the child to his manman. And all will be made as right as it can be made."

Anaïs shifted on her feet. She cast a quick glance up at Mambo Johanne, who shrugged helplessly.

Anaïs turned her attention back to Lucille, who had craned her neck back to watch the dancers swirl around them. Lucille's feet bounced back and forth and she hopped on her bottom.

"We shall do as the Lwa ask," Anaïs said. When Lucille grinned at her, Anaïs continued, "Can you take me to the jé-rouge? To the child?"

Lucille nodded eagerly and reached into the Judge's jacket. She pulled out the white handkerchief and unfolded it to reveal the shriveled umbilical cord. She sniffed. She squinted. She sniffed again.

Suddenly she leapt to her feet, arms swinging. The empty piman bottle that she still held in one hand narrowly missed Mambo Johanne's head as the priestess ducked out of the way.

"This way, this way," she sang. Lucille followed the celebrants on one more circuit around the tree, then headed out onto the street.

Anaïs quickly moved after her, hissing as she stubbed a bare toe on the stone steps. She ran to catch up with Lucille, only pausing for a moment when Mambo Johanne called out her name. She turned, reaching up just in time to catch her shoes before they smacked her in the face.

"Bondye ale avèk ou!" the Mambo yelled, the blessing almost lost beneath the laughter and general noise of the crowded street.

Anaïs waved a hand over her shoulder, trotting until she was beside Lucille again. She awkwardly pulled on her socks and shoes while Lucille skipped, singing a happy nonsense song about mice in the sugarcane.

Fireworks cracked overhead, flashing red, blue, and green. Electric streetlights gave everything a soft golden-orange sheen, in contrast to the sparking red-gold of torches carried by celebrants. The crowds were densest near the city center and around the larger hounfo. Fortunately, Lucille led her west, towards the ocean. The masses of people began to thin out. The fine buildings and houses of the city slowly gave way to smaller homes and businesses, fisheries and canneries, brothels and drinking holes. The smell of the ocean grew stronger and soon Anaïs could hear the waves over the sounds of celebration.

Lucille suddenly stopped at a crossroads, in the shadow cast by a fish market. The street to either side of them ran parallel with the ocean. Ahead, the road continued for a little ways, then disappeared; the ground turned rocky and uneven, with tough grasses sprouting here and there. An old building huddled at the end of the nonexistent street, leaning heavily to the side. It was probably within reach of the waves at high tide.

Lucille pointed with the empty bottle, and Anaïs peered through the darkness.

A stuffed caiman hung from the porch, swinging in the breeze.

Not, in itself, alarming.

But still...

"I have to go now," Lucille said. She hopped on her toes.

Anaïs dragged her eyes away from the old buildings. "Thank you, Lucille. Please tell the Baron and Manman that we will complete the Work they have set us."

"I know. Just watch out for the rat."

The body of Celestin twisted and jerked. A loud gasp. Anaïs caught him as he fell, but he was too heavy and he dragged her to the ground. She grimaced as her elbows banged against the stone of the street, and the little bottles and daggers inside her jacket dug into her chest.

Celestin panted and shivered as he leaned against her.

She held him for long minutes, listening to the ocean and watching the old building where the caiman swung in the wind.

The Judge finally pushed himself upright. He grimaced at the empty bottle and set it aside. He started to speak, coughed, and tried again.

Anaïs handed him a little bottle of spring water. He guzzled it, and then a second bottle. When she would have handed him a third, he waved it away. "Save it," he grunted, voice rough. "Who was the gede?"

"A little girl named Lucille. The jé-rouge and the child are in there." She pointed her chin towards the old building. "Baron wants us to kill the monster and the fallen mambo, and... expose the thief." She shrugged. "Return the boy to his mother. I have my blessed daggers, fresh basil, and spring water."

The Judge seemed to consider her words for a moment, then nodded, feeling around inside his own pockets. "My cane, my spectacles, basil, water, powdered caiman skin, dried wasps, graveyard dirt." He handed her two small bottles each of dirt and dried wasps. "I wish Lucille had not drunk all of the piman." He was silent for a few minutes, gaze narrowed as he studied the rundown building. "We will circle around. Come in through the water. Once we have it trapped..." He lifted his cane. "You see the child..."

"Put myself between him and the jé-rouge."

The Judge held out his hand, and she helped him stand. He settled his glasses on his nose. "Let's get to Work."

* * *

They backtracked around the block and came out a street up from the old building. Moving between shadows, they crossed to the rocky shoreline and stepped into the water. Further and further out they walked, the water swirling and tugging at their legs, making a wide path as they moved around the old building and then back towards it from the sea.

The water rose to Anaïs' knees, her hips, above her waist. She lifted her jacket to keep its precious contents dry.

Finally, they reached the shore again. The old building with its stuffed caiman stood directly ahead. Port-au-Prince lay beyond it, all stone and wood and soft golden lights and laughter and singing, the crescent moon a sharp arc against the fireworks.

As they neared the building, crouched low, Anaïs drew four daggers from her pockets, two tucked into each hand. When Celestin motioned for her to move to the left, she nodded silently. Dripping salt water, her steps careful on the rocks and sand, she stopped when she reached the porch. Testing the wood with her foot, she stepped up, pausing as it groaned low.

Nothing. No sound, no movement from within.

Another step, and another, closer to the back door.

Celestin moved into her peripheral vision.

The wood creaked.

They both stilled.

A rumbling, guttural growl.

Celestin rushed forward and kicked the door. It crunched. Another kick and it slammed open, banging against the wall. He ran forward, pausing just across the threshold, head rotating as he surveyed the small room.

He pointed his cane towards the far dark corner, reaching into his jacket and flinging a bottle of spring water as he did so.

Anaïs flicked a slim little blessed dagger from each hand.

The bottle shattered, and the daggers thunked into a solid mass.

There was an abrupt, hard-edged howl. Red eyes appeared in the darkness and then a shape, and it moved, the daggers glinting where they poked out of it. It flowed across the room, the human that it possessed contorting and twisting and leaping in such a way that muscles should have shredded and bones should have popped. It moved to the wall, all shadows and claws and red eyes.

Another bottle of spring water, soaking it to reveal more of its true form.

Then fresh basil, the green leaves sizzling as they clung to it.

Anaïs moved around the Judge, further into the room. Her eyes darted. There. Tucked tight in the corner, a blanket drawn up around his little body as a futile shield: the child, his eyes closed tight, his arms over his head.

She lunged forward, rolling, and came up facing the room, her back to the little boy.

The jé-rouge hiss-howled, swiping at her with its claws. Anaïs leaned back, feeling them catch on her jacket and tear through the fabric. The Judge was shouting. Another bottle of basil leaves shattered against the creature's side. It spun and crawled up the wall towards the ceiling. Its claws dug into the wood, shadowy fur bristling.

This time, Anaïs threw a bottle of wasps along with her dagger. The knife cut into the shadows, slicing part of it away, revealing more of the human form beneath. The glass bottle shattered, loosing the dried wasps. They clung to the jé-rouge, digging through its fur, stinging, stinging, stinging again.

One last bottle of spring water from the Judge, and then a bottle of graveyard dirt.

The jé-rouge twisted, shrieked, and fell. It slammed into the floor, claws ripping.

The human form within the shadow was now fully visible: an old man, his skin heavily lined with age, his fingers knotted with arthritis. His eyes were closed in sleep, his mouth slack.

Celestin cracked his bottle of graveyard dirt against the floor immediately in front of the monster. The soil spread out in a fine dust, driving the creature towards the wall. He hastily planted the tip of his cane in the dirt and began to draw the veve for Baron Samedi.

Anaïs threw her own bottle, dusting the wall behind the jé-rouge.

The monster crouched and panted, spinning on its feet.

Trapped.

It snarled, the sound growing more frantic and desperate as Celestin lifted his cane. He pulled off the outer casing, revealing the sword inside.

He pointed the sword down at the ground and then at the jé-rouge. It glinted in the light that reflected in through the door.

"Your grave has already been dug," the Judge said.

He slid forward, arm extending. He drove the sword through the chest of the jé-rouge.

The monster writhed, whimpered. The shadow began to evaporate, leaking away. It floated into the air, a little at a first, then more and more, like it was being sucked away by a hurricane. More and more shadow whirled into the air. The eyes were last, the red streaking, swirling...

And then, nothing.

The old man sprawled against the wall. His breath stuttered unevenly.

There was a tear in his shirt, but otherwise the sword had left no mark on him.

Anaïs wiped sweat from her forehead with the back of her hand. She was shaking and breathing hard.

The Judge stepped in front of her, sword still free. "Did it draw blood?"

Anaïs felt her clothing. The jacket was shredded across the front, just below her collar bones. The shirt beneath... untouched. She unbuttoned it just in case, running the palm of her hand over her upper chest.

The skin was smooth.

She sighed with relief. "Non," she said, shaking her head.

There was a small movement behind her.

She turned on her toes, smiling down at the little boy. He was perhaps four years old, his skin a rich sepia, his eyes a dark green, his hair black and straight. A true child of Haiti, born of Africa and Europe and Taino.

She held out her hands to him, whispering nonsense comfort words. Then she started to sing Lucille's silly song about the mouse in the sugarcane.

The boy smiled, shy. He dropped his arms, carefully crawled out from under the blanket, and snuggled against her chest.

"Come along, yon ti kras." She stood, holding him tight. His little fingers curled in her shirt. "Let's take you home to your mother."

"Mmm," the Judge said. "And to the right one, I hope."

* * *

They covered the old man with a blanket and left him unconscious with an unopened bottle of spring water. There was no time to deal with him right then. They would tend to him later, or send Mambo Johanne.

They moved as quickly as they could through the streets. Three blocks, four, five. A wagon finally appeared, shallow barrels that would have been filled with rum now rattling empty in the back.

Celestin waved down the driver, the horses whuffing in annoyance as he pulled back on the reins. After loudly objecting, and complaining that he was tired and was missing his own independence party, the driver finally agreed to turn around and drive them back to the Palais du Gouvernement; Celestin had to offer him one free service of his asking, though. ("My son-in-law is yon moun sòt. Who knows what trouble he will get into. I will have need of a Judge some day. I will. I know it.")

Anaïs perched on the front seat next to the driver, the boy in her lap. Celestin sat in the back, his legs dangling over the edge. She continued to sing, but also worked in a few questions, such as the boy's name and where he lived and the name of his manman.

He answered none of them, just smiling shyly and giggling.

Fighting their way through still crowded streets, they finally reached the wrought iron gates of the Palais du Gouvernement. Anaïs had no idea what time it was, but, judging by the height of the crescent moon, it had to be nearing midnight.

When the Garde saw her and Celestin, they waved the wagon through the entrance.

The driver perked up as they drew closer to the Parlement Haïtien. He sat straight in his seat and smoothed his shirt.

"What a story I will have to tell the family!" he crowed.

Kapitènn Massez stepped forward as the wagon drew to a stop, Travieso at her side. As Anaïs climbed down, the boy braced against her hip, Travieso pushed past the Kapitènn.

He crossed his arms, chin thrust out, but stopped out of arm's reach. Anaïs felt her lips tug into a satisfied smile at that.

"And who is this?" Travieso demanded, glaring at the boy.

Celestin twirled his cane, spectacles on the bridge of his nose. "We are about to find out." He paused, grinning, and tilted his head. "Why don't you come with us?"

Travieso gurgled. "I – that—"

"Kapitènn Massez, if you would please?"

"Wi." The Kapitènn closed a hand around Travieso's arm and steered him around towards the great mahogany doors of the Chamber of Deputies. The electric lights to either side glowed yellow, but curtains were still drawn across the windows.

"Kapitènn," the Judge said, voice low, "please leave the creature and Chevry to us. When the time comes, your concern should only be getting the child to safety."

Massez was silent for a moment, then nodded. "I am trusting you to keep our Présidente and our République safe."

"Your faith is not misplaced," Anaïs assured her. She shifted the boy in her arms, reaching into her torn jacket to pull out a pair of daggers. The boy sniffled, his head against her shoulder as he drifted closer to sleep. As they reached the steps, Anaïs tilted his face into the side of her neck, hoping to keep that sleep free of at least a few nightmares.

Celestin banged his cane against the door and then pushed it open. The hinges squeaked.

The interior was completely dark, save for two stuttering electric lights. They cast a pallid glow across opposite walls of the room, the light just barely reaching the edges of Chevry's dress.

She was still sprawled across the floor, her leg still crooked at a wrong angle. Her face was lined with pain, but she held her head high.

The once-human creature loomed behind her, hand gripping Chevry's hair. Rats crawled through its hair and a caiman snapped at them from near its ankles.

Baka chattered at them from the thick shadows and there was the sound of wings flapping and beaks snapping high near the ceiling.

Travieso spun and tried to run, his eyes wide.

Massez tightened her grip on his arm, using his momentum to jerk him back around. She shoved him in front of her, pistol in her free hand.

"This is absurd!" he exclaimed, voice rough. "There is no reason for me to be here!"

The once-human creature snarled, the baka echoing the sound. It filled the room, almost crawling across Anaïs' skin. The boy twisted in her arms, and she gently pressed his head back down again. When he stilled, she lowered her hand, daggers at the ready.

"You have brought him. My boy. My sweet child."

"No," the Judge said.

Another snarl, higher pitched this time. Baka crawled out of the shadows towards him, bright-eyed rats, sharp-toothed caiman, and wriggling snakes. A boar shook its tusks at Anaïs.

"This child is not yours, not by blood or bond," Celestin continued. "A rotted thing such as you could never have given birth to him. He was given to you – sold to you – by one who had no right to do so. This is La République d'Haïti, and no one may be sold to another, especially not a child."

He continued further into the room, cane clacking. When Massez would have followed, her hand still gripping Travieso, Anaïs gave a small shake of her head. The Kapitènn lifted an eyebrow, but remained close to the door.

"So who was it?" Celestin asked. "Mm? Who stole this child from his rightful manman and sold him to you? And why, mm?" He tilted his head down. "I thought, perhaps, you, mon Présidente. A gift to win all of... this." He waved a hand at Chevry's dress and the room and the building as a whole.

Chevry's eyebrows twitched, but she remained silent. The thing's grip on her hair tightened, pulling back her head and arching her neck.

"Non." Celestin shook his head. "You have the taint of baka about you, but it is fresh." The Judge lifted his cane, but did not turn around, his gaze fixed on the creature. "But on you, Monsieur Travieso, it is an old stain. It has seeped into your skin, your brain, your soul. I am Baron's Judge, and *I see you*."

Travieso panted, sweat running down his face and neck.

Anaïs saw Chevry's eyes dart to him, to Celestin, and back again. Understanding, then rage and grief, clouded her features.

"That is why you came here, wi?" Celestin asked the once-human thing. "He got you the child once before, surely he could retrieve the boy from this jé-rouge that stole him away from you in the night. Another trade, eh? Chevry for the boy? Well, there he is!"

The Judge waved his cane over his shoulder. Anaïs shifted on the balls of her feet, adjusting the boy's weight in her arms.

Celestin leaned towards the once-human thing, his legs carefully braced.

"He was never yours to begin with. And he shall never be yours again."

The thing leapt.

It streaked through the air, clawed hands raised. Sharp teeth in a rotting mouth, dirty hair streaming behind it—

—the baka rushed forward.

Anaïs pivoted, tossing the child to Massez. The Kapitènn released Travieso, caught the boy one-armed, and dashed for the door. Her pistol cracked as she took aim at a boar, and then a snake, and then a caiman.

Travieso shrieked and ran after her.

Anaïs caught a glimpse of Garde rushing into the Chamber, pistols and rifles at the ready.

"Close the door!" she yelled, and turned away, making for the Judge.

Her daggers flitted through the air, sinking deep into the rotten flesh of snakes and rats. The blessings of Manman Brijit ate away at the baka, reducing them to piles of ash. Crows and bats swooped for her head. She caught a crow with her bare hands, slammed it against the floor, and drove a dagger through its head.

Pistols and rifles boomed, filling the Chamber with the stink of gunpowder. Baka shrieked and hissed, writhing on the ground as they were pinned by bayonets and knives. The Garde yelled, advancing.

Closer. Anaïs made her way across the room, one eye on the baka, one on the Judge as he wrestled with the inhuman thing, his cane braced against its chest. It bit at him, jaws snapping a fraction of an inch from his throat. Its clawed hands tore at his jacket.

They rolled, Celestin ending up on top. The thing kicked at him. Bottles of basil, graveyard dirt, dried caiman skin, dried wasps, and spring water fell out of his shredded coat, shattering against the creature and on the wooden floor. It screamed louder, rolling in the herbs and water and broken glass.

A rat tumbled out of the thing's hair, running up the cane and up Celestin's arm.

Anaïs caught it with a dagger, sending it sprawling across the floor.

The Judge slammed his cane across the thing's face once, and then again, and again. It sagged against the floor.

He pushed himself to his knees and ripped the outer casing off his sword.

"Your grave has already been dug," he said. It was not his voice alone, though, but also the voice of the Baron. They echoed one another, digging into Anaïs' bones, into her skull.

The Judge drove his sword down, through the thing's heart and into the floor beneath. The once-human creature screamed and twisted, raking its claws through his jacket, his shirt, into his flesh. It smoked and began to turn a greyish-black. Skin and hair and torn clothing disintegrated, collapsing inward. It crumbled, turned to ash, and was gone.

A rifle boomed again, and then silence.

Anaïs dropped to her knees in front of Celestin, her chest heaving. Over his shoulder, she saw Présidente Chevry sag in relief. Anaïs looked around, finding only more piles of ash where the baka had been. Dust drifted down around her from high above, sticking to her hair and clothes.

The Judge sank back on his heels, sword across his lap. He smiled tiredly at Anaïs, his clothes in tatters. His spectacles tumbled off the bridge of his nose, bouncing on their purple ribbon. "It is rarely clean Work, mon apprentie, but it is always necessary."

* * *

Chevry was somewhere inside the Presidential residence, a small army of doctors seeing to her injuries. Other doctors tended to injured Garde, and to the Judge, propped up in a chair on the front portico. When they came to Anaïs, she waved them away with a smile and a murmured thank you.

When they finished with Celestin, his torso and arms wrapped in thick bandages, she sat down beside him. Flags snapped in the breeze above them and the sweet smell of the pots of flowers drove away the stink of gunpowder.

"Travieso?" the Judge asked. He rubbed at his bloodshot eyes.

"Gone. But not for long, I imagine. Massez is hunting him."

"Feh!" Celestin snorted. "And the child?"

"Inside. Présidente Chevry is taking responsibility for him until his parents or other family can be located."

"She already has been."

They both stood as Chevry stepped onto the portico. She wore a new dress, though a much simpler one, her sash of office across her chest. Bandages wrapped around one arm and ointment had been applied to the bruises on her face and neck. She leaned heavily on a cane, limping as she approached them.

With her other hand, she led the little boy. He smiled at them tiredly.

A pair of Garde followed her at a respectful distance.

Chevry carefully lowered herself into a chair, motioning for Anaïs and the Judge to do the same. She tugged the boy into her lap and kissed his head. "Etienne. I named him before he was born. And it is my fault that he wound up with that... thing." She paused, drawing herself up straight. "I had little support, initially, when I entered politics. I earned more, over the years, but never considered the Presidency. Then I became pregnant." She smoothed a hand down Etienne's cheek. "Travieso was there, at the birth. He told me that my son died. I was lost in my grief. Travieso persuaded me to focus on my work, on the Presidency. And I won – quite unexpectedly. Now I know why."

"He made a deal," Anaïs whispered. "He traded your child for evil maji provided by that thing. Power for you meant power for him – at least a taste of what he could never have."

"Wi." Chevry looked away for a moment. "I cannot trust that my victory was truly the will of the people. Nor can I trust my own judgement, if I listened to someone like him." She kissed her son's forehead. "I will be stepping down."

Celestin nodded sadly, but in agreement.

When Chevry looked at Anaïs, she mirrored her mentor's nod.

As they sat there in tired, thoughtful silence, watching Etienne drift to sleep in his manman's arms, Anaïs heard a low chattering. Turning her head, she caught a glimpse – barely a glimpse – of a rat skittering across the portico, down the steps, and into the Haitian night.

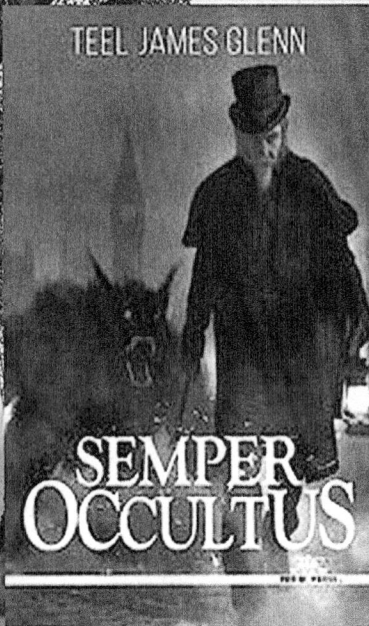

COMMITTEE OF MYSTERY

ROBERT GUFFEY

"This is the name we call ourselves by on occasion, eye to eye, at the end of one of those afternoons when we can no longer find anything to divide us. The number of secret doors within us keeps us most favorably disposed, but the alert is very seldom given."
—André Breton, "Soluble Fish," 1924

The Committee of Mystery meets every Monday night in the basement of a fortified compound. The house is situated on top of a hill overlooking the coast. Several locals have told me it looks like a haunted house. That's because it is. It's haunted by ghosts.

Several of them.

Earlier tonight, there were twelve of us at the meeting. Sometimes there are more, sometimes less. We sat around the long wooden table dressed in our uniforms: a wide brim hat; a burlap mask with fangs painted around the mouth; eyeholes that curved upward like the feline, painted eyes of an Egyptian princess; long black capes with multicolored fringes (the fringe depended on your station within the Committee); a black dress shirt; black dress pants; sturdy black work boots. Our belts were outfitted with ropes and knives and other necessary tools.

We didn't know what any of us looked like underneath the masks. Our voices were muffled by the burlap, and our names were codes made up by someone much higher in the Committee. Our group was not the only one, not by any means. Rolling committee meetings occurred throughout the year, all over the country.

All over the world.

* * *

I arrived at the studio around one in the afternoon on the thirteenth of October. The purpose of the show was to stir up controversy among several different guests, all low-level celebrities. Like me. They taped several shows in a single afternoon, then aired them – one per day – about a week later.

Tonight's theme on *The Snake Pit* was, incredibly, Crop Circles and Alien Abductions. Jesus H. Hadn't we debunked those old chestnuts back in 1987? These bugaboos are like unsinkable rubber ducks. You shoot them down, then inevitably they'll rise out of the earth again like ol' Vlad the Impaler come back from the dead. No amount of truth can keep a bad idea down, as long as that bad idea is infused with *mystery*. Everybody loves mysteries, and people find them where they don't even exist. Fortunately, that makes *our* job a lot easier. As P.T. Barnum once said, "No one ever lost money underestimating the intelligence of the American public."

The host, Guy Ludwig, introduced me with his usual smug demeanor. As always, I nodded at the camera once, curtly, without smiling. It's best to maintain a poker face when you appear on these shows. It's very much like a game, after all. You don't want your opponent to know what you're thinking, what your real agenda is. Don't wear your brain on your sleeve.

My opponents, on the other hand, were a husband and wife team who had just published a book through HarperCollins about their year spent aboard a Pleiadian spacecraft. If not for the fact that the pair had already published a number of bestselling novels, the book would never have been released by such a huge outfit. For the most part, UFOs are a dead subject in the New York publishing circles. It was alive only in the form of occasional documentaries on the History Channel or the National Geographic Channel or on late night AM talk radio. Because I'm the editor of *Skeptical Times Magazine*, many of my readers credit *me* with helping to bring about the subject's demise. (I see no reason to dissuade them from this belief.)

These two lovebirds were clearly bohos, upwardly mobile Bohemians. They haunted New Age conferences at night while tending to their debentures and their 401Ks and their tax shelters during the day. Their clothes were very mundane, not flashy at all. That was a good choice on their part. The audience would identify with them. As always, I made sure I wore my uber-academic's outfit: a wine-colored corduroy jacket, beige dress pants, brown loafers, tan socks decorated with the symbol of Pi. I even had a scarf around my neck and a pipe in the left breast pocket of my jacket that I never intended to smoke. Why not? I liked to project a standoffish air. It was all part of the plan.

* * *

I felt liberated underneath the burlap mask. Free to be myself. No one else on the Committee knew who I was, of course. They knew nothing about my connection to *Skeptical Times Magazine*, but not one of them would disapprove if they did know. The magazine served our purposes. The more my magazine debunked the paranormal, the closer the Committee grew to completing its goal – a goal that might never be completed. But that didn't matter. Attaining the goal wasn't the point at all.

The Comte de Nox, Deputy Wraith of the Committee, placed his hand over the wrinkled map laid out across the table. His hands were covered in a black cloth painted with the image of a skeletal hand like that of a child's Halloween costume. His whole suit was done up that way. It might sound silly, unless you see the thing coming at you down a dark alley. Or in your bedroom at night, when you're alone, and your parents are a thousand miles away.

The Comte de Nox pointed at one particular area of the map outlined by a hasty circle drawn in red ink. The location was perfect, a ranch in the middle of nowhere. Only a small family lived there. The Legion of Phantasms had surreptitiously entered the premises earlier in the day and removed all the bullets from the firearms.

The preliminary work had been done. Now it was time for us, the Hordes of Anubis, to take over.

* * *

Ludwig started out by asking a question of the lovebirds: "Mr and Mrs Ravineau, I should state right now that I'm an open-minded person. I read your book and was fascinated by it. I'm not closed to the possibility that alien life forms are visiting Earth. If life exists on other planets, why couldn't they come here? That seems natural to me. But in your book you say you spent a year on this alien spacecraft. How is that possible? According to everybody who knows you, according to the police and the FBI, you never went missing for an entire year – not even for a few weeks."

"Time works differently on their spaceship," said Mr Ravineau. "These people can bend time and space to their will. And that's what they are, too. *People.*"

"That's right," said Mrs Ravineau, "we don't like to call them 'aliens.' It's such a negative term and just engenders fear. We like to call them 'people' because that's what they are. They're just people who happen to live on another planet. They have the same emotions and desires you or I would."

"Well, speak for yourself," Ludwig said, and the studio audience laughed. "I don't have any particular desire to kidnap anybody. That's considered a crime on this planet. Why would they bother? What could they learn from you two, except how to save money on nice clothes?"

The crowd's laugher swelled.

"Yes, it's true, we like to make our own clothing," said Mr Ravineau. "It helps the economy. If everybody did that, we wouldn't have any need for sweat shops in India."

"Did the aliens tell you to make your own clothes?"

"Yes, they did," said Mrs Ravineau.

"Yeah?" said Ludwig. "Did they tell you it wasn't 1967 anymore? Did they accidentally drop you into Greenwich Village during the Lyndon Johnston Administration or something?"

The audience laughed again. Whenever Ludwig manifested one of his trademarked smirks, they would immediately burst into guffaws like Pavlov's dogs. It didn't really matter what he was saying.

The couple took it in stride. "No," said Mr Ravineau, "they wouldn't make a mistake like that. You see, they exist out*side* time. The past, present and future are all the same to them. They take note of time only as it relates to *us*."

"So why are they interested in you?" Ludwig's smirk had faded, indicating he was ready to get back down to business again.

"Because," said Mrs Ravineau, "for generations these people have been trying to jumpstart their dying biology by mixing their genes with ours. They're creating hybrids in order to perpetuate their race. Earth women are giving birth to their children and vice-versa."

"Have *you* had one of their children?" Ludwig asked, one eyebrow raised.

"Four of them," said Mrs Ravineau.

* * *

We parked our trucks far away from the location, then rode our horses to the ranch. Our steeds were as black as night and specially trained to be as silent as the ghosts of ghosts. Half the team remained behind and set up the boards and chains in order to begin inscribing the massive circle in the field. Tonight we had decided to create a tetrahedron circumscribed inside a sphere. We always tried to outdo ourselves.

The other half of the team surrounded the house, then began the assault. My new Lieutenant, a Goblin Prince of the Veiled Realm, proved himself worthy of his position by leading the assault inside. He slipped through the open window, which was still unlatched due to the careful planning of the Phantasms, then began crawling across the wooden floor, hissing at the man and woman in the bed. The Goblin Prince was dressed like a half-snake, half-human creature from an alternate dimension only tenuously related to ours. The young woman woke up screaming. The old man grabbed his rifle, aimed wildly, pulled the trigger. Nothing happened.

* * *

The audience burst into guffaws all over again.

Ludwig mugged for the camera. He let the laughter swell for awhile, then at last said, "You know, I don't think I'm even gonna try to come up with a punch line for that one." The audience laughed even more. For some reason, the audience thought Ludwig was at his funniest when he acknowledged something was humorous but didn't bother to make a joke about it. I had noticed this pattern throughout all the shows I had been on. The absence of jokes drew the most laughter. I always thought that was strange.

Ludwig turned to me. "So, professor, what do you think about all this? I bet you think it's all just a buncha poppycock, right?"

Ludwig insisted on calling me "professor," even though I wasn't one. I didn't complain. It served my purposes. "Yes, poppycock. That's a very good word to use, actually, maybe more accurate than even you know, Guy. First," I said, reaching under my chair to pull out a manila folder, "allow me to comment on the word poppycock itself. The word poppy is a reference to the flowers from which opium, or heroin, is derived. You know, drug-talk, the ramblings of someone under the influence of a narcotic. And cock is, well…" I let the sentence trail off. The audience laughed. And I didn't even bother to make a joke. I'd picked up a few techniques from Ludwig.

Ludwig arched his left eyebrow for the camera. "Are we all getting screwed by heroin addicts right now, is that what you're saying?" The laughter swelled again.

"Not exactly heroin addicts," I said, tossing the manila folder onto the small wooden table that sat between the four of us. *LSD addicts.* I knew, of course, that it was impossible to be physically addicted to LSD, but I also knew that the majority of the audience would not be aware of that, so I used their ignorance to my advantage. I pointed at the folder. "Right there are the arrest records in New York City for the two lovebirds over here. They were drug traffickers. They both went to jail for it."

* * *

By that time the Goblin Prince had crawled back outside.

Meanwhile, the Feral Imp of the Dark Brigade darted through the back door, kicked open the door to the children's room, then launched into one of the prayer-shrieks of Mamuzeda. What was it like seeing that little man, only about four feet tall, dressed in the hairy skins of a dozen different dead animals? They only caught a glimpse of the thing before it was gone, the enigma retreating under the cover of night.

I saved the final task for myself. Before the final syllable of Mamuzeda's prayer had been uttered, I scaled up the side of the house with the help of the spikes in the bottom of my boots and the spool of rope I'd strapped to my back. I stood on the wooden roof, the cool wind blowing my cape behind me like a magic carpet. The gibbous moon hung low in the sky. I wish the Committee's official photographer could have accompanied us on that particular raid. I'm certain I must have struck a romantic figure there in the darkness, silhouetted by the moon.

I kneeled down beside the sunroof the family had built in their living room. I flipped on the holographic projector connected to my belt and beamed the hallucinatory images through the roof. Mamuzeda's prayer had brought the parents running into the hallway where they had a perfect view of the living room. In my ear I heard a soft voice say, "They see it! Now they're headed towards the kids' room..."

From the parent's perspective, it would look like demonic, ephemeral shadows were dancing in the dark sanctity of their home. Dancing and laughing and beckoning them forward. Beckoning them to dance as well? What would they think, these vacationing materialists from Beverly Hills? That their world had turned upside down?

* * *

The two lovebirds looked astonished. So did Ludwig. I hadn't told him I was going to do this. My magazine has several private detectives on the writing staff. It wasn't difficult to track the records down. It was simply that nobody had ever bothered to do it. People are lazy, especially book publishers.

"That information alone invalidates their entire story," I said. "How can you believe anything that comes out of the mouths of acidheads?" I smirked, crossed one leg over the other, and settled back in my chair.

The lovebirds didn't know what to say at first.

Ludwig turned toward them with a somber expression on his face. "So is it true?" he asked.

For a moment there I thought Mr Ravineau was going to launch into a real Ralph Kramden moment: "*Homina-homina-homina-homina.*"

Instead he gathered his wits together and said, "This is no secret! We wrote about LSD in our first novel. We've talked about our experimentation with hallucinogens in several interviews – in *New Dawn Magazine*, for example. LSD does not induce hallucinations. It allows you to see reality as it really is."

The audience laughed. Mr Ravineau looked nervous.

Mrs Ravineau said, "You see, at an evolutionary level, we are prevented from seeing the miraculous. The true visionaries, the ones born with the talent of second sight, they often die very young before they have a chance to reproduce. That quality in the brain that gives them their paranormal abilities are therefore not passed down the line. The people from the stars who seeded us here planned it that way. We're part of a grand, on-going experiment. They want to see if we can attain the transcendent despite all these obstacles built into us biologically. What will prevail, the miraculous or the merely biological? That's the real question at stake here, not arrest records."

Ludwig picked up the file. "Are these records real?"

"Oh, they're real," I said, taking a sip from my coffee mug. E=MC2 was scrawled all over the mug.

"I haven't looked at them," she said.

"Were you ever arrested for drug trafficking?"

Mrs Ravineau laughed. "That was the subject of our first novel. Of course. We've written about this at length. They were trumped up charges."

"Not according to the police," I said.

Mr Ravineau laughed bitterly. "And you believe everything the NYPD says? I thought you were a skeptic."

"Occam's Razor would dictate that I trust the NYPD's word over the testimony of a couple of latter day Timothy Leary acidheads convicted for selling hallucinogens to children."

"Children?" said Mr Ravinaeu. He slammed his fist onto the table. His wife tried to calm him down, but it did no good. "They were narcotics agents, sent in to entrap us!"

"But you *thought* they were children."

"Should one's age be a barrier to one's spiritual *enlightenment*? Some truths are greater than the laws of this country, sir. You'll come to realize that one of these days."

I just spread my hands and smiled at Ludwig. "I don't think I need to say anything more, do I?"

"I don't think so," Ludwig said. "We don't have time to say anything more anyway. So we'll be right back after this. Oh, and please shut off your critical faculties *during* the commercials."

* * *

Only forty seconds had passed. That was enough.

I switched off the Demon Light and scurried back down the side of the house. Again, the voice spoke in my ear. "Master, they hear you coming down the side of the wall. They're running into the kids' room. The kids are crying, screaming. The parents are trying to calm them down."

"They won't be able to," I whispered to myself, as well as to my deputy.

"Those tykes will be traumatized for life."

"Sometimes trauma's good for you," I said. "Trauma can change your life. Give you a purpose."

"Or it can just drive you mad," said the Deputy Wraith.

"I guess we'll see, won't we?" I dropped to the ground, reeled the rope back in, tossed it over my shoulder, and darted off across the grass with the Goblin Prince and the Feral Imp of the Dark Brigade running at my heels. To the casual observer, we would have looked like proud, imposing shades sent out by Lucifer to disturb the sleep of the just. Ah, yes.

I flicked on the Demon Light as I ran, casting Doré-like hallucinations before me. We leaped on the backs of our steeds who still waited patiently for our return. We rode into the night. From the windows of the ranch, it no doubt looked like the guardians of Hell itself haunted the lonely fields.

The Wraith Doctors had finished with all the animals by the time we reached the stables on the outskirts of the farmland. They had been slaughtered, their uteruses surgically removed. The Deputy Wraith and his expert staff, agile men with nimble fingers and a love for the knife and the thread, had performed the entire job within mere minutes. There were thirteen men on the staff, all dressed in black surgical coats. The Deputy Wraith stood amongst them, tall and proud, a living skeleton lording it over the dark disciples of the Hippocratic Oath – the *true* Oath, the one they don't teach in medical school. Death and his Doctors of Pain.

Death said, "Our job is finished here. And you?" It was strange to hear Death's voice echoing in my ear.

"We're done." I turned to one of the Doctors and said, "No extra souvenirs this time. Just the uteruses." They nodded silently, pale moonlight glinting off their round eyeglasses, their little black bags dangling from impossibly thin, bony, spidery fingers.

* * *

It was rather uncomfortable during the commercial break. The two lovebirds shot up out of their seat and began yelling at Ludwig about being sandbagged. The producer came over and patiently explained the waiver they signed before coming on the air. There were some warnings of a lawsuit, but I knew that was ridiculous. There could be no lawsuit because the Ravineaus were right. I wasn't revealing any secrets. They had talked about their arrest quite extensively in the *New Dawn* interview. That's how the private detective found about it. Once we knew about the arrest, it wasn't difficult to get the records from the NYPD. How could we get sued for 'revealing' something they'd already talked about?

Of course, the beauty of it all was that it need not be a huge revelation. I knew full well that not one person in Ludwig's audience read *New* freakin' *Dawn*, so it would *seem* like a revelation to them. And appearance was everything on television.

Appearance was everything, period.

* * *

All of us converged on the outer edge of the field and were pleased to see that the Spectres of Geometry had completed the tetrahedron. It would no doubt look impressive and precise from thousands of feet in the air. The Spectres were professionals. They never failed in their appointed tasks. Never.

We piled into the trucks and drove off at a safe and sane pace. The compound was our next destination. We would then return to our respective abodes, sanguine in the knowledge that we had performed yet another in a long line of good deeds.

I steered the truck, and the Deputy Wraith sat beside me. Yes, Death was my passenger. Always at my side. Good ol' Death. So much mystery arises from the absence of Death. It felt so comforting, therefore, to always have him at my side.

All of us were experts at what we did. Our work was important, and we knew that. We were esoteric psychologists, treating the ennui of a world too sophisticated for itself, bored with its own technological progress. The world needs *mystery* to survive, and our committee provides that throughout the

Earth. We always have, always will. As inoculations. Vaccinations against complacency. Against those who think the human brain capable of explaining all phenomena through an opaque lens of rationality.

We need only wait now, I thought, wait for the medium of gossip and rumor to spread undiluted *mystery* through the channels of AM radio waves and word-of-mouth and internet chatter like a viral meme.

I sat in silence that night, watching sharp dark clouds cut across the surface of the gibbous moon – the milky white eye of a blind, cyclopean god. An image that once inspired wonder in man. Yes, ancient humans used to believe that the moon could infect a man with madness. But as the centuries crawl by, even the moon has been raped by too much knowledge. Werewolves have become extinct. And vampires are relegated to comic books and jokes on late night television.

Humans are addicted to mystery. Shorn of a steady diet, they begin to withdraw... into themselves and their petty, mundane problems. People have to believe that something greater than themselves exists beneath the surface of their everyday realities.

Even if it's not true. Even if it's not true.

The spirits in the attic? The flying saucers twirling through the night like falling stars? The angels aflame in the cathedrals and hospices? The gargoyles creeping through the tombs of dead saints? That was us. That was the Committee.

That was me.

That was me.

That was me.

Blessed with membership in the inner sanctum of these mysteries, I'm also cursed with the knowledge of their true source. I'm forever cut off from the allure, the erotic call of ignorance. The only way I can experience mystery is by dressing up in its clothes and emulating its moves, its distinct tone of voice. Always I'm off-key, but at least I try. And more often than not, the impersonation is successful. So few people have experienced mystery, how can they say I'm not the real thing? No longer are there ordained experts in mystery, not like there used to be. Those who profess to be such, those who appear on TV talk shows and attempt to debunk the otherworldly, merely serve to push more and more people toward the Other Side. Back into the dark ages. Back into a life of dread. Into the panicked dawn of a race that feared the unpredictable wrath of the moon.

Alas, I no longer fear the moon. Nor do I fear the night, nor Death itself. I fear instead the looming day when the Committee of Mystery is discovered, our cathedrals decimated, our members burned at the stake and our sacred mission wiped out by the light of rationality. That will be an uncertain day for us all, for we Masters of Mystery would sooner release the werewolves inside us all than see the moon destroyed entirely by sunlight.

* * *

Ludwig was a little pissed with me. He said I should've cleared my revelations with him first. I told him there was nothing to worry about, my lawyers had already checked everything out. He seemed to take that as a definitive answer. He himself was stymied by my professorial air. I think he looked up to me. Ludwig was the kind of fellow who always wanted to be smart, but never quite had the chops for it. He can utter clever little bon mots, but that's about it. He's a middlebrow who desperately wants to be recognized as far more than that. That's where his oft-mentioned political aspirations come into play. A run for the Governor's office isn't going to work out for him, though. Ludwig tries too hard and it shows. People distrust that kind of thing. I try too hard myself. In fact, I go out of my *way* to try too hard. But then again, my imperative is somewhat different than Ludwig's. I don't want to be liked. Quite the opposite, in fact.

By the time we came back from the commercial the Ravineaus had calmed down. We finished out the next segment politely enough, but it was too late. Too late for me, that is. On TV I came across as an academic bully and the Ravineaus came off as the underdogs, the romantic rebels. Ultimately, by the end of the show, the sympathies of the audience leaned toward the Ravineaus. The sales of their book went through the roof. Which is exactly how I planned it. It's exactly how we planned it.

The Ravineaus should thank us. Fortunately, they've never heard of us.

Nobody has.

And, if we have our way, nobody ever will.

* * *

I write these words for my own benefit, no one else's. Nobody will ever read them, for I have sworn an oath to never reveal the secrets of my sacred body. In order for there to be mystery in the world, there must first be something worth hiding.

Any mystery worth keeping is worth keeping in the dark.

From the streets of Los Angeles to the beaches of Miami
to the cafes of Paris and beyond...

Desperate people can be found anywhere.
These are their lives and their struggles.

GUILTY CRIME STORY MAGAZINE

is on Amazon in Kindle and Print

JAMES BOJACIUK

The Omega Factor: Festival of Darkness
Written By: Natasha Gerson
Produced & Script Edited By: Xanna Eve Chown
Read By: Louise Jameson

Students die after mediumistic experiences. Strange, clownish figures stalk the streets. The history of British occultism looms. *The Omega Factor: Festival of Darkness* refuses to slow down.

If you have read this column for any length of time, you knew to expect this review. *The Omega Factor* audios have become one of my favourite occult detective series. The series has proven consistently exceptional. Some of the best, deepest characters and a uniquely grounded approach to both the occult and the idea of the government backed investigative body. The occult threats never leave the parapsychological behind. Budgets and run-down facilities are more of a threat than telepaths. The focus on the genuine makes the breaks from reality all the more striking.

Since the third series, *The Omega Factor* audios have continued with novels and audiobooks. Primarily set in the television incarnation's 1970s, they significantly develop the original cast as deeply as the modern-set audios have developed Dr Anne Reynolds.

All of which takes us to Natasha Gerson's *Festival of Darkness*. The daughter of the series creator, Gerson also played the mute Morag in the original series. In the third audio series, she wrote the great '*Let Us Play*', so I was curious to see what she could do with the space afforded by a novel. With the room to explore things more fully, she truly impresses.

The characters are exceptionally handled. Dr Anne Reynolds has become

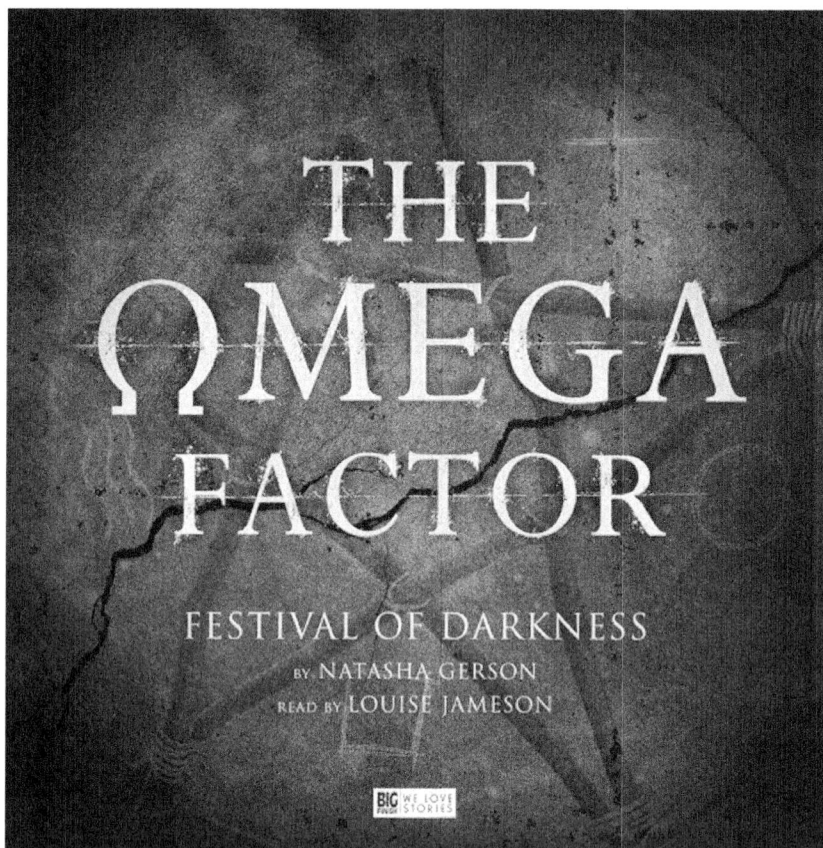

one of my favourite characters, and she is given quite a lot to do in this novel. We see the woman she would later become, and she begins to develop in that direction. Roy Martindale is handled uncommonly well. He is, of course, a prideful, pompous fool. The chapters in his perspective are delightful, and use this to their advantage. Unlike the chapters from Anne or Tom's perspective, there is very little objective description and nearly no dialogue. Instead, these brief chapters are a guided tour of his ego. As prideful monologues, impressions of the wider world only wash in vaguely.

Speaking of the way Gerson approaches her work, one facet of this novel particularly stands out. *Festival of Darkness* is largely written in the present tense. It never feels awkward or unnatural, a rare feat. Experienced novelists often stumble here, but Gerson pulls it off and makes it look easy.

Gerson perfectly marries the two halves of occult detection: the preternatural and the mysterious. The preternatural sequences and threat are genuinely frightening. Visceral horror gets all the press. If a horror movie makes the news for its content, or a book is talked about in hushed tones, you can rest assured that it features gore and abuse. But for sheer power and fear, *mundane* horror strikes the hardest – which is to say, slight expressions of horror built from the world around us. The classic pool sequence in Val Lewton's Cat People (1942) is powerful precisely for how understated and grounded it is. Something is in the room. The character is in a vulnerable position. We can hear that *something* and see its shadow. This has stood the test of time for fear for nearly 70 years, when so much competing visceral horror has been forgotten.

All of this to say, Festival of Darkness does this admirably well. The little thrills of terror have stuck with me in a way a cleaver through the head has not. A scene as simple as a *thing* trying to pull open a locked car door is depicted so clearly, and with such reality, that it clings to the audience. The majority of the horror fits firmly in this tradition. Sudden fires, psychological affects, and a burning theatre are all drawn so well. It is intimately familiar, and therefore immediately pictured; because it is immediately pictured, we feel a stark fear.

In this way, it's firmly in the tradition of Charles Williams and his 'supernatural shocker' thriller novels of the 1930s and 1940s. The magic is grounded, slight, and hard to find. Much of it is drawn from the tradition of British occultism. Powerless or only slightly sensitive characters combat incredible dark powers. All of the horror is of this mundane sort – such as the doppelganger from *Descent into Hell* (1937) or the possibly imaginary meeting of magic from *War in Heaven* (1930) – and all the richer for it.

Festival of Darkness is an unusually effective, grounded horror story.

Let us turn to the audio-specific aspects of this production. Previously I praised Neil Gardner and Tanja Glittenberg for the exceptional sound design in the Carnacki the Ghost-Finder audiobooks. This, however, is a much more pure production. It is simply the reader and us.

We are fortunate to be in such safe hands as Louise Jameson. She is an intimate, enjoyable reader. She works overtime to pull the listener in. All characters receive their own voices, but it never becomes awkward or distracting. It is more than a reading, it is a sustained performance.

Like conventional mysteries, one of the most difficult feats is to make an occult detective story novel length. Mysteries seem naturally made for short stories. Horror naturally leans toward the shorter. *Festival of Darkness* handles these difficulties of genre with ease. A case just expansive enough to fill a novel without becoming belaboured or messy, with chances for both genuine detective work – deduction, physical examination, clues – and horror.

Festival of Darkness is excellent and well worth your time. One of the best recent works of occult detection, and simply an impressive horror novel in its own right.

The Scarifyers: The Nazad Conspiracy
Written & Directed By: Simon Barnard
Starring: Nicholas Courtney as Inspector Lionheart; Terry Molloy as Professor Dunning; David Benson as General Warlock and Aleister Crowley; Cicely Giddings as Lady Walsingham; Pasha Kanevsky as Doctor Lazavert; Seva Novgorodtsev as Dimitri Romanov; Owen Oldroyd as Chief Inspector Fang; Stuart Silver as Doctor Slither and Pickering; Alexei Voronkov as Lieutenant Sukhotin; Zinoviy Zinnik as Bobo the Magnificent
Music By: Edwin Sykes

1936 is ending with a bang. Russian demons strike out, killing men in strange ways; Rasputin's prophesied resurrection approaches; Aleister Crowley takes issue with horror fiction.

Recently, *The Scarifyers* transitioned to being a podcast. Now that you may listen for free – at the rate of one episode per week – there is no better time to check it out. I'd like to give the first serial attention in its own review: *The Nazad Conspiracy*, by Simon Bernard.

Marrying *The Adventures of Tintin*, the works of M.R. James (particularly the adaptation *Night/Curse of the Demon*), and Universal Horror, *The Scarifyers* is a unique series. Horror, action-adventure, mystery, and comedy are brought together in an engaging package.

The foundation to a good series is the characters. As M.H. Norris is so fond of saying, *People come for the premise, but stay for the characters*. The premise sells itself well. "A professor who writes ghost stories and a sceptical police officer team up to stop preternatural threats" is an interesting, unique spin on the occult detective team. Two likely mystery stars are put together into infinitely strange mysteries.

But we stay because of the characters. I've remained engaged with this series thanks to them. I haven't stayed for *the professor who writes ghost stories*, but for Professor Edward Dunning, a confused, happy, mouse of a man who does his best to remain positive in terrifying situations. I haven't stayed for *the sceptical police officer*, but for Detective Inspector Lionheart, a brave, sardonic man who refuses to let the world stand in the way of justice.

Dunning and Lionheart are more than excellent characters on paper. They are made greater through the performances of Terry Malloy and

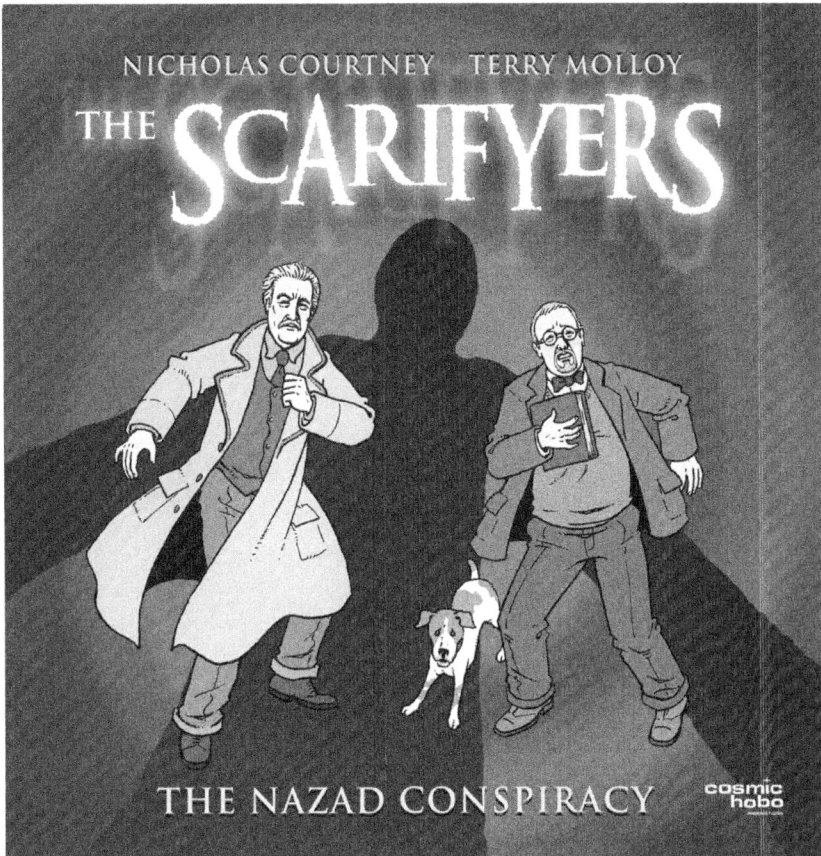

NICHOLAS COURTNEY   TERRY MOLLOY

THE SCARIFYERS

THE NAZAD CONSPIRACY   cosmic hobo

Nicholas Courtney. Malloy makes his mouse into a real man, whose every sentence reveals himself; he is an open book. Courtney gives Lionheart a guarded depth, but we feel his decades on the force. The magic happens when they share a scene. Their chemistry and interplay is the foundation of those early episodes, and pull the listener in. The best comparison is *The Thin Man* films. Fast, engaging dialogue rules the day. A sample exchange:

Dunning: "You gave my audience a terrible fright."

Lionheart: "I thought that was your job."

Characters can't exist in a void, which takes us to our case. Russians are dying in strange ways. A man drowns with no water nearby. Another man receives a fatal dose of cyanide – despite being locked in a cell with no visitors. Soon Dunning and Lionheart – one through witnessing a Russian's death, the other through being assigned the case – are thrown into the depths of preternatural crime. A circus, an aristocratic club, separate arms of Russian survivors, and the resurrection of Rasputin are only some of the puzzle pieces they're forced to work with.

Genre mashups are difficult to balance. Comedy-horror is difficult enough, with only a few works effectively handling that balance (chief among them, *Abbott and Costello Meet Frankenstein*), but manageable. Comedy-action/adventure-horror is a much more difficult blend to brew. Most works attempting this combination falter. Fewer still work when mystery is sprinkled in as well. *The Scarifyers* is one of the very few to balance these competing concerns, and still deliver effective, engaging stories. Simon Bernard and, later, Paul Morris handle this perfectly.

The Scarifyers: The Nazad Conspiracy is a great story from an excellent series, and well, well worth listening to. One of the longest running occult detective audio series, it has proven its longevity is founded in its quality. I can't recommend it enough, and all the more now that the price is free. Find *The Scarifyers* through your podcast access of choice (Apple, Stitcher, etc.), and enjoy.

BECOMING ART DECO

CRISTINA L. WHITE

My name is Solas Bierman. I am a registered psychic investigator with the Leon Institute of Psychic Phenomena at Damari. The cases I have worked on over these last eleven years are well documented at the Institute, and the records are available to legal entities and verified researchers in the field.

I have degrees in art history and ethical design. Because of this background, and through natural inclination, many of my cases are art-related. With permission from the principals involved, I have collected certain of these cases in my new book, *Missing and Found in Art*. By way of sparking interest before the book is released in the spring, my publisher has decided to make available this advance copy of my most recent and, in many respects, most unusual case.

Briefly, the facts are these: two years ago, I was contacted by the architect Armand Barsini. He asked me to investigate the mysterious appearance of a sculpture in the entryway of Lowe's Hotel in South Beach, the Art Deco district of Miami. The Florida hotel is well-known both to international visitors and American vacationers.

The hotel had extended its grounds and Barsini was contracted to design the structures in those expanded areas. He had included in his designs a space for a large sculpture as part of the fountain which is the central feature of the curving driveway up to the grand hotel. But a dispute with the artist originally commissioned for the piece led to a delay, and finally, Barsini decided to start fresh. He was in the process of reviewing submissions from three well-known sculptors when, overnight, the bronze figure of a seated woman appeared in the space designated for the proposed sculpture.

The figure is slightly larger than life-size, and matches the proportions the architect had intended in the overall design. There was no one, either on the staff or among the hotel guests, who could recall seeing the sculpture brought to the site or installed.

The hotel doormen who were first aware of the sculpture reported that they were at their usual station at the main entrance of the hotel in the early hours just before dawn. They recalled talking casually with one another, and then greeting a fellow staff member who was just ending her shift. As she exited the hotel, she commented on the sunrise sky. When they looked up and out, they saw the sculpture, in place at the fountain. It was soon determined that the figure had not simply been set down at the site. It was embedded in the stone, as if it had always been there.

Barsini did not wish to replace the sculpture – he was, in fact, enthralled with it, and declared at once that it was perfect in every detail. He felt as if someone had reached into the deep recesses of his mind and created the ideal piece for the fountain.

He had heard of my work and contacted me because he wanted to know how the sculpture had been brought to the site and embedded there, and because he had a sincere desire to credit the artist. It was his hope that I would be able to psychically connect with the woman who had served as the model for the piece, and that this would lead him to the sculptor.

To those who work in this and related fields, my methods are well-known, and my approach to this case did not differ significantly from other investigations I have conducted. Upon arrival at the site, I walked around the entire area, staying aware and open to whatever might be there. I felt moved to sit beside the sculpture, and to take the same position as the figure of the seated woman: both feet on the ground, the upper body leaning slightly forward, the forearms resting on the upper thighs. I waited for any images that might come, for a scent or taste, for any words or sounds. I let my eyes relax, and almost close, because the woman's eyes are slightly open; it is impossible to determine if she is at that moment about to close her eyes or is just awakening. I then had an image of myself writing. I always have with me a pen and a notebook. I opened the notebook and began writing.

The entire transmittal came to me in this way, in the form of automatic writing. It was recorded in one sitting shortly after I arrived at the site. The original transcript is both date and time stamped, recorded as a transmittal from a woman called Nicole.

I received this transmittal several days before the Wolfsonian Museum in South Beach found that twenty-four black and white photographs of Lowe's Hotel had been bequeathed to the museum by the eccentric collector Hans Ziegler. Those photographs of the hotel exterior are in the style of the noted photographer Nicole Germain. Additionally, Hans Ziegler stated in his bequest that the camera containing the film from which the photographs were made was found on the grounds of Lowe's Hotel. As to why he did not tell anyone about finding the camera, we can never know. Julia Warren, Nicole Germain's lover and life companion, confirmed that the camera was of the same make and model that Ms Germain had with her on the day she disappeared.

I also wish to emphasize that the transcript was recorded and complete nearly two weeks before I was contacted by Ms Warren. She had been staying at a mountain retreat when she learned of the sculpture, the photographs, and my investigation.

Twenty-seven years ago, in 2004, Ms Warren and Ms Germain vacationed in South Beach. They stayed at a hotel on the edge of town. On their last full day there, Ms Germain took her camera and some pocket money and went on her own for one last trip to the central area of South Beach. She never returned.

The mystery of her disappearance has yet to be solved. I cannot say that my investigation of the sculpture at Lowe's Hotel settled the matter. My transcript of the transmittal from a woman called Nicole was, for the most part, scoffed at by local officials. However, after studying photos of Nicole Germain, it is impossible not to see the resemblance between Ms Germain and the fountain sculpture at Lowe's.

Upon reading the transcript, Ms Warren was at a loss to understand how the details of the arrival at the airport could so closely match her arrival with Ms Germain at the Miami Airport. She also felt a shock of recognition in the description of how she and Ms Germain had met, many years ago, in the city of Santa Monica in California. And she was struck by the 'voice' of the woman in the transcript, who referred to the 'Palace' in San Francisco – Nicole Germain's term for the Palace of the Legion of Honor. This woman was also completely captivated by South Beach, as Nicole Germain had been.

Ms Warren stated that Nicole Germain had for many years kept a journal. She recorded the sights and events of their travels, as well as her own inner experience of the places they visited. She thought that she might eventually create a travel book that combined some of her journal entries with her photographic work.

The transformation described in this account presents a mystery unlike any other I have encountered. I have no explanations to offer. I can only present the transmittal as received, and leave you to surmise what you will, in the case I call *Becoming Art Deco*.

Solas Bierman
November 18, 2031

What follows is a word-for-word record of the transmittal from a woman called Nicole, received through automatic writing.

* * *

Even while we were up in the clouds, before the plane touched down, I knew I would like this city. Maybe it was the sparkling blue green water down below, or the sound of Spanish being spoken everywhere around us, the easy talk of Cubans returning home to Miami. On the long walk through the terminal, the first scent was of cinnamon. The air was thick with spice. We passed a stand where young women were selling cinnamon buns, and I found it was enough to breathe in the scent; I could taste their sweetness without even touching them.

On our way to the baggage claim area, we passed a Cuban restaurant. I lingered in the entryway, entranced by the atmosphere of the interior. I wanted to settle into the restaurant's underwater light, the array of bottles at the bar, the milky greens and blues, the lilac and pink glass. I imagined walking toward the slender bartender and sitting at the bar. The bartender would lean forward and ask what I'd like to drink, and I would order something tall and frothy in an ice-cold glass. Julia would join me and say, the same for me. On the second round we would move to a table, and order food. We would eat, and be happy, and a little drunk.

None of that happened, except in my mind. Instead I waited with the luggage and watched the people around me while Julia stepped outside to have a cigarette. When she came back, Julia said, "Nicole – go out and feel the air."

I stepped out into the steamy, tropical air, mixed with the intense heat from the cars, vans and taxis. It would be all right, I thought, once we were away from the traffic. I could sense the real atmosphere that existed beyond the airport, like a heartbeat I had yet to hear; I felt its pulse.

Were there symbolic signs at every turn, or was I simply looking for them? Perhaps there was some meaning in the fact that the lady we rode with in the van was on her way to a psychic healing conference. Later, as I stood on the balcony of the hotel and felt the soft ocean air fill my being with sweetness, I would look back and think yes, there was meaning, a hint of the healing this place would provide.

It all unfolded like a dream. Standing on the terrace of the hotel, I looked along the arched columns out to the sea, and then turned to see the buildings across the way – deep coral, sea foam green, Tuscan yellow. The colors washed over me like balm.

It was everything here: the pastels mixed with deeper Mediterranean hues, the palm trees, the languages – French, Haitian, Spanish – the Art Deco everywhere, in the architecture, the interiors, the furnishings and objects and small details. It was the gorgeous kite flying over the beach, the scent of the Atlantic Ocean, the sun. But most of all, it was the air, air that seemed to kiss you, gently caress your skin, fill your lungs and every cell of your body with a Dionysian elixir.

From the moment I arrived in South Beach, I felt this place embrace me, and draw me deeper into a passion for Art Deco that had begun in the spring, in San Francisco.

* * *

I came to Art Deco late in life. Although it seems more accurate to say that it came to me, in the form of the Art Deco Exhibit at the Palace of the Legion of Honor. The museum is now commonly known as the Legion of Honor, but I have lived in San Francisco for a long time – I prefer its original name, and I've taken to calling it the Palace.

I went to the exhibit for the first time soon after it opened, and fell in love with what I saw. I knew that I would return there, not just once more, but many times. Here, in one place, was the look, the feel, the color and line I had been responding to all my life without realizing it was all of one period.

I was born just as Art Deco came to a close, born in the midst of a war that spanned oceans and continents, a world focused on the battle between fascism and freedom. Now, the decades had gone by, and the curators and collectors, the lovers of Art Deco art and design, could gather all the best and put it on display. For some it mattered not at all, for others it was a morning or an afternoon spent viewing the art and artifacts of another era. For them, though they might admire the collection, it would probably be forgotten.

But there were others, like me, who felt a deeper kinship, an accord with this era that resonated down to the soul. During my visits to the Palace, I found people who banded together to keep the Art Deco era alive. They were people who had gone to put on a costume and discovered their wardrobe.

I found these people, this era, just as my dark hair was turning to silver. Maybe that's how it was for me, to find true love late in life. I had the same experience with Julia. I was nearly forty when I drove along a boulevard in Santa Monica, on my way to meet friends for dinner. I saw Julia in front of the restaurant, dressed in long, flowing shirt and soft slacks, standing beside a pale blue balustrade. I knew her at once, knew this was the love of my life.

Julia always wanted to give me whatever I wanted. This year, I told her I wanted South Beach. The exhibit at the Palace led me to look at my own city with a different perspective – I could see with new eyes the Art Deco that was still there in San Francisco, mixed in with all the other periods, to be discovered by the faithful. At the same time, I spent the spring and early summer living like an addict, returning to the Palace again and again for a fix, a pure concentrate of Deco design. As autumn approached, I told Julia I wanted to go south to Florida, to a place devoted to preserving the period. In South Beach, for a handful of days, I felt I could immerse myself in Art Deco.

* * *

We stayed five miles north of South Beach itself. The plan was to have easy access, without being subject to the intense scene of the clubs, the crowds, all the people who wanted to be part of the mix in the district known as Sobe. I didn't know what to expect here. I only knew I hadn't expected – or even anticipated – this air, these skies, the purity of the sea and sand. It was a revelation.

The first morning there, I walked across the terrace of the hotel, past the sapphire blue pool, down six steps, and out to the beach front. I stepped onto the sand – warm, white, deep. There was white sand as far as the eye could see, in both directions, glistening white sand. A turquoise sea and bright blue sky, with stacked white clouds that the wind carried overhead and across the horizon, great, snowy domes that changed constantly, sumptuous, breath-taking.

That morning, and all during the week, whenever I was on the beach, I felt slightly unreal, as if I had been temporarily caught in a photo shoot for a glossy magazine. Nothing, I felt, could be this perfect, and be real.

The first day there, Julia wanted to relax and recover from the trip. But I was excited and impatient. In the heat of mid-afternoon, I went on my own to the heart of South Beach. I started at Fifth Avenue and walked north along Ocean Drive. Like the beach that looked as if it were in a romantic movie, the main promenade lived up to all I had heard about it. The café tables on the terraces and sidewalks were filled to capacity. Music spilled out onto the street, people strolled by, crowding each other, glad to be there, happy to see and be seen. Most of them were young, many of them endowed with the beauty only youth can give. And the Victor, the Bentley, the Colony, the Imperial and the Waldorf Towers – all the lovely Art Deco hotels presided over this scene with the perfect assurance that they were the real stars, the magnet that drew all these people together to a place where the genteel buildings graciously provided them with a splendid setting.

Even then, in the first hours there, I began to think about staying. I imagined life in an apartment on one of the quiet side streets, in a place where Julia could read for hours on end and swim to her heart's content in a warm pool. I would wake early and go out to photograph this endlessly beautiful world in the morning light. In the afternoon, I would return to our home. I would stay indoors during the mid-day heat, and re-emerge in the evening to be part of the music and the dancing, to feel the cool sea breeze from the Atlantic on my skin and bask in the glow of Art Deco neon.

But the fantasy only led me to think of the reality. Would we really live here? Would I be content to die here? Even as a child, I had considered death. Now, as Julia and I grew old together, I wondered who would be the first to go. If Julia went before me, how would I cope? Julia was my home, my shelter and warmth, the magnetic center that gave me the freedom to roam and explore, to seek out and photograph the beauty glimpsed in certain faces, the fall of water on stone, the bend of light on a summer hillside.

Without Julia, I would be cut loose to stumble into the one death I feared, still drawing breath while my body began to crumble. Still alive, but unable to hold my own against the inevitable crush of age.

I wanted to choose my death. I wanted to die dancing, moving in rhythm to the beat and sound, to dance furiously until my heart burst from love and I fell to the floor and was gone. Cause of death: ecstatic exhaustion.

Yes, that would be all right. I didn't want a long life – at least, not as a human being. To be a glorious work of art, marveled at down through the ages – that would be a good long life. Or to be one of those beautiful South Beach buildings, preserved, admired, lived in and loved by Art Deco addicts like me. Then, let the decades and centuries wear me down; I would be content.

The days flowed by, languorous, easy vacation days that were slipping away too quickly. I didn't want it to end, but time was relentless. There was nothing to do but live the moments as they came, and be grateful for this time here. I wrote postcards to my friends, trying to describe it: "... a narrow strip of paradise, elegantly encumbered by Art Deco architecture, embraced by the azure ocean to the east, the ultra-blue Bay of Biscayne to the west. Exquisite morning light... Latin jazz nights, endless white sand beaches..." How could they possibly know, without being here, what it was like, what I was feeling?

I loved being in the constant presence of this Art Deco world, feeling its line and color permeate my thoughts, my dreams. And in the same way that I felt connected to the aesthetics of the period, I felt a kinship with this stretch of coast. Each day, as I walked on the quiet beach outside of our hotel, I felt a deep sense of peace pervade my spirit. With the vast canvas of sea and sky around and above me, all the joy and trouble of the constructed world dissolved. The beat of the urban tempo was still present, but it was dimmed, like the skyscrapers of Miami in the distance, ethereal against the sky during the day, jeweled towers of light in the evening.

On our last afternoon, a bank of clouds blew in. For the first time in a week, there was a gray sky and cool air. Julia wanted to stay at the hotel. She wanted to finish reading the novel she was engrossed in, and to pack and prepare for the morning flight. I wanted to go one last time to the central district of South Beach. There were still streets I hadn't seen, places I hadn't been able to photograph.

As I had on the first day here, I started at Fifth. But this time, I headed west, away from the clubs and hotels, the cafes and restaurants, away from the trade that mainly catered to tourists. I walked along the residential tree-lined streets, past the houses and small apartment buildings of South Beach. All these places where people lived were painted in evocative pastels – coral, lilac, sage, pale yellow, ivory, pearl gray and topaz, as if a child had been put in charge, and this town was her coloring book. Lush magenta trails of bougainvillea hung over the gateways, ferns and palm trees swayed in the afternoon breeze.

After a while, it began to rain. I walked in the rain until it started to pour, then I ran across the wide span of Washington Avenue and stood in the entryway of the Wolfsonian Museum. I looked up at the museum banner and smiled at the words there: We Have So Much To Show You.

Yes, Wolfsonian, you do. Yes, South Beach, you do. The Art Deco feast.

Finally, it stopped raining, and I continued along the avenue. At twilight, I reached Lowe's Hotel. I loaded my Tri-X, my last roll of film, and shot it all, rapidly, before I lost the light and the splendor of this place. I heard the shutter clicking, the whir of the film advancing, I felt the thrill of images captured, held for a moment – this, now this, and this. Beauty. I am beauty, behold.

When it was over, I pressed the tiny rewind button and wound the film back. I left the cylinder of film in the camera, and walked across the curving driveway to the island fountain that welcomed the prosperous travelers arriving here in luxurious cars. I sat down there, on the wide rim of the fountain. I set the camera down beside me, and felt my body settle. I was tired, and content.

I closed my eyes, and listened to the voices along the street, the hum of traffic in the distance. I felt the onset of evening. I seemed to float outside my body, and I saw myself sitting there, but I was not myself. I had expanded, become larger than life. I was someone transformed by South Beach, a woman of the nineteen-thirties, my body shaped by handsome, clean lines, a body that was heavy, strong, powerful. I was this new world where anything and everything was possible. I was an Art Deco sculpture, supremely elegant.

I had soaked myself in a rainstorm of beauty, breathed it in, become saturated with color. I was newly blended, newly made. As I started to open my eyes, I knew I had become my true self, the soul of Art Deco. I had become this place.

End of transmittal.

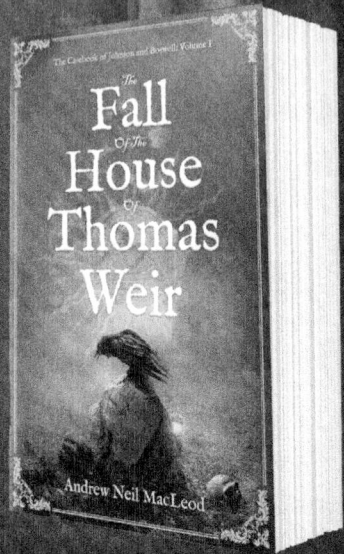

SPIRIT COUNSELLOR

UCHECHUKWU NWAKA

My world is empty. Like a void. I have long stopped seeing the meaning in many things. In the mundane normalcy of life. Every day, I wade through the stream of the collective consciousness called society. It does not welcome me. No. Let me rephrase. I've forgotten what it feels like to be welcomed. I just want to exist. Somewhere, but away from here. Anywhere but here. The other day I saw a place offering out-of-body experiences for just two thousand naira. Pathetic. Fortunately for everybody involved, it's a scam. There was no spirit presence there at all.

Maybe that's why I tried it.

The office is the same as it has been since I started using it a year ago. A shelf filled with books I'll never read gathers dust in the corner. It has to look that way so that my clients can at least take solace in my supposed authenticity. The table is ridden with termite holes, but I don't mind. There's a part of the ceiling that's leaky, close to the wall, and the wall paint on that side has peeled from a dull yellow to a duller brown. To think this is the most colourful piece of my life.

Sometimes I recline in the cheap chair and let my inner voices yell me to sleep. Their words are never nice. They're always taunting me. Telling me about things that might have been. Who I could have become. I used to believe I had great mental fortitude. I still do, I think. The only reason the voices can't get me depressed over the inadequacy of my life is because I'm already depressed about the inadequacy of... oh my.

Today the client is a part-time student of poly-Ibadan. She's crying and shaking and I tell her it's going to be all right. I don't believe that, and I know she doesn't either. However I have to tell her this because I believe I would have turned out differently if someone had told me the same words. If only those who were more interested in how much school work I had missed had asked how I really felt inside. I don't know where they are now, and still don't know how I feel inside anyway. That's why I say I feel like nothing. Life is just a tedium that wears me out, yet I am not capable of ending it myself. At least this young girl did not encounter the Eater. Merely a plot from her father's side. A plot involving carved likenesses of Eastern deities and blood sacrifices that has left her motherless...

69

I dread evenings. Especially ones like this when it rains and I can hear the wallowing of the elements. At these times, I feel a different kind of melancholy. One born of languor, but fundamentally heavier. I remember that I am a slave to the shackles of everyday existence. Of dealing with the normal. Occasionally I replay my tapes. Listen to what people described as their encounters with the preternatural. There's nobody to listen to mine, and I tell myself I don't care. Before the Eater there was nobody, why should there be one after? Maybe listening to others will help me overcome this gaping absence in my chest. This chasm widening to my own self-destruction. I walk home under the rain, my umbrella barely surviving the downpour. I lie on my bed and wonder why I can't do things differently. Then I wonder about happiness, and all its different yet elusive colours until I fall asleep, and dream about that night when everything was taken away from me.

Then the cycle starts again.

* * *

This week begins, as other weeks, uneventfully. Some UI students come by, spinning tall tales just to mock me. I'm all too accustomed to this, what with my office located beside their school. Their exuberance is the height of my disgust. Their ignorance! I don't let it show though. I scribble into my notes and give them advice that I never got to utilize in uni. I can see that they are disappointed when they leave and I wonder what they thought when they saw 'Spirit Counsellor' on the sign post beside the post office.

A man comes by later. Surprisingly he is considering a money ritual and the babalaawo has demanded he kill his mother. I laugh, but not audibly, or visibly. That's some Nollywood bull-crap if ever I've heard any, but who am I to judge? Besides, the gods of this land are as mad as they get, compared to anywhere else in the world. Trust me, I know.

I'm tired of listening. Once upon a time I wanted to be a doctor. Wore a pristine lab coat in the university barely two plots away. Even though there was always pressure as a first child and all that, I seriously believed I could make it work. Even though people's expectations of me were disproportionate to reality, I felt I could make it happen if I worked hard. Working hard was my strong suit. Truly. Med school was hard, yes, but there were geniuses who made it look doable. If only my parents understood that some things couldn't be achieved with only hard work. Some talents could never be surpassed with countless hours of mind-wracking study. Why did you want me to be the best student so bad? I believed I could too, and that was my mistake. I should never have relied on my ability to work hard so badly. If I hadn't given in to that final despair of failure, I might not have ended up here... even though now, I can't really see what I could have become.

I tell the man to stay true to his heart. Crap advice, but he'd do whatever he liked anyhow.

I complain a lot about how my life is like a mosquito's breeding ground, but in the madness of freedom that is the world today, I take solace in my routine. In the assurance of where I'd be and when. I am happy that I can be unhappy with my stagnation. That bleakness comes and goes and is with me anytime I want it. I am content with the nothing, and how I do not want to change it. Even if I do not get what I want – info on the Eater – through counselling, I can afford to work and feel others' sadness at the same time.

I am still hoping to someday meet another victim of the Eater.

I close my door and step out. No customers today. I check my phone but I don't really have anyone to talk to so I stash it back. *Quebes* looks tempting across the road, but I'm on a tight budget – spirit counselling doesn't pay much. I give the plaza's gatekeeper an ambivalent greeting and enter UI's busy highway.

I'm a partial seer. Once in a while I can see spirits. A side effect from surviving the Eater and its carnage. I don't fixate my gaze on them, that's folly. Telling them you're aware of their presence. One time last year I had an attack... horrible spirits of Yorubaland. I had just returned to Ibadan, and had to seek help from a very shady babalaawo. Let's call him Baba.

All experience, I tell myself.

When I get back to the office, there's a girl standing before my door. Another UI student. I'm not really in the mood to tolerate 'smart' youths, but I resign to charge her extra for any crap she tries to pull.

"Come in."

I notice her eyes are red. Swollen. Her gaze is distant, far away. Her hands are shaking where they clutch her bag, and I'm starting to get worried.

"Hey, are you okay?"

She sits on the couch... and doesn't make a snarky comment. Okay. Her braids are the black kind, her nails painted black, but everything else looks off. Like she didn't remember to complement her already existing looks. By this, I mean makeup and extremely flashy jewelry. Not that I'm one to talk. I'm already questionable enough running a practice at twenty-something and looking like a forty year old man deep in mid-life crisis.

"I'm sorry sir," she says after a while. "Somebody recommended this place."

Really? Someone actually saw the online ad?

"Okay," I'm going by the book here. "I'm Dominic Uchenna." I consider adding, "Call me Nna," but Nigerians don't necessarily follow those rules.

I opt for, "What do I call you?"

"I'm Ifunanya Akinnola," she managed. "I'm a UI student."

"So. Do you have a problem?"

"Um, yes..." she pauses. "You're going to call me crazy."

"You're sitting in a room rented by a self-proclaimed spirit counsellor. Crazy has already started."

"Sir..." her voice drops down an octave. "I can see spirits."

Not entirely interesting, and I haven't totally thrown out my theory of her lying, but I want to see how much she knows. Or thinks she does.

"It's nothing spectacular sir," she continues. "I've been able to see them for a long time actually. Even spirit related things. Cursed people, haunted houses. Things like that."

Okay. Now my interest is really piqued. She's a seer, yet she looks like she's seen a ghost.

I let her talk. That's my job.

"Sir I'm not crazy."

"I don't believe you are," I reply. "But you look pretty shaken. Why?"

"Sir I saw something..." and her voice breaks. "It... it ate my friends."

Dread washes down my spine, and I hope she doesn't hear the quiver in my voice when I ask, "What do you mean?"

"On the way to our hostel, three of us. There's this construction going on in school and they've been felling trees. Last night... Well there was this grandfather tree on the way to our hostel and it was taken down. It was hollow inside."

"A hollow grandfather tree?"

"Yes. It... it was inside. It ate them."

"And how did you get away."

She fishes inside her bag. After a brief moment she reveals a pendant. "My grandmother gave this to me before she died." Ifunanya hands it over, and I run it over my fingers. "I think it's a protection amulet. I woke up on my bed so I thought it was perhaps a dream... until their roommates declared them missing."

"They aren't the first ones no?" I ask, recalling a missing girls story I happened to hear the last set of UI students talking about.

"There have been rumours on Hive that people were disappearing, but there were no confirmed reports and I thought it was merely a prank by Tech boys."

Hive? That was the name of the club I attended those years ago. Where everything happened.

"Ifunanya. What is this Hive?"

She pulls her phone from her jeans. "Here. It's a social platform by some group of people in the Faculty of Technology." The app looks like a blend of WhatsApp, Twitter and Nairaland. However the interface is a bee-yellow, and the icon is a curious cowrie.

Same as that place.

"... they are trying to make it the next Facebook. It's currently UI exclusive, so it only works on UI grounds."

This girl is on to something. I can feel it. This is what I've been waiting for.

"I'd like to see this app work," I say. "And this tree."

I can see the terror in her eyes, and I kneel before her on the couch, holding her hands. "I can help you get closure Ifunanya. Don't worry."

Worry!

"Okay Mr Uchenna."

"Mr Nna is just fine."

I have her wait outside while I push the shelf back. There's a bunch of occult stuff in its back compartment, but I don't need all of them. Just my wand – yes, it's an actual wand, but not the Harry Potter nonsense – and a chalk I stole from Baba's shrine. If his god is truly potent, let it fight the spirit for him.

I take a vial of holy water too... for good measure.

Evening descends quickly. We are on a path inside the school, but I know it's not leading towards the dorms, private or otherwise. The silence is deafening, and even the obnoxious crickets are wise enough to stay away. Ifunanya is on her phone now, and a chat group is texting feverishly about the two missing girls from two nights ago.

"That's the tree, Mr Nna."

I let my torchlight illuminate the tree. Somehow the path has led to a quiet and secluded part of the street. Further up, lights from teachers' quarters spill into the street on two sides, but both sides of the road where we now stand remain dark. The street lights are broken, and vegetation grows where the houses are absent. The darkness whistles an eerie tune, and the moonless night is not helping.

The fallen tree is truly hollow inside. The wood resembles wrinkled skin. Tough and scaly. I feel my heart beat faster in its chamber. Fear. I try to relish it. For the first time in four years I'm truly alive. I run my fingers along the length of the trunk while Ifunanya holds another torchlight in her hand, a safe distance away on the road. Truthfully, nothing seems out of the ordinary. The area is spooky, but that's just it. I'm about to chide myself when I notice a trail of finger marks. It's… red.

"Ifunanya," and I get my phone to take a picture. Only, the spot on the tree is coming up stainless on the camera. Ifunanya trudges into the bushes. She's literally trembling. She can see the blood stains too, but her camera can't. A spirit phenomenon? A strange thought occurs to me then. Hive has a camera feature, and I ask her to try it.

Chills run down my spine when the blood stains on the tree trunk register on its camera.

"Oh my God."

"What exactly is this app? How come you and your friends ended up here?"

Her hands are trembling now. "I-I don't know sir," she pauses. "We were actually supposed to meet someone here. Our hostels would have been locked, it was past midnight. We wanted to stay in a friend's apartment at a lecturer's boys' quarters. It was at Bini Road, and we didn't know where that was."

"So this friend sent you their location. On this app?"

"Y-yes."

"This is messed up Ifunanya. Is there anything else you're not telling me?"

"No. No sir."

I run my fingers through my hair. We leave the bushes. It's not safe.

"So your hostel closes by ten. How did you manage to get locked out by midnight? A party?"

She nods.

"Very well. But why this app?"

"Didn't I mention?" She swipes down her notification panel. "The app doesn't need data. Not always. I ran out, so I had to use this."

Did I mention I was getting a bad feeling?

We're under one of the street lights now, and Ifunanya's frightened expression is stark illuminated under the yellow light. I'm looking through the app, but I can't figure out how it sends messages without data.

"What's your friend's address?"

She reads it from her phone. Block 4 BQ. Bini Road.

We are on Amina Way. After a few wrong turns and some helpful directions from a delivery guy, we're in front of the house. Luckily, the girlfriend is outside. Probably coming in for the night.

"Tolu!"

"My God, Ifunanya!" They hug, and do girl stuff. Her expression darkens when she notices me.

"You called the police?"

"I'm not police," I say. "I'm Dominic Uchenna. A spirit counsellor."

Tolu casts Ifunanya a strange look. I'm quick at ad-lib. "I just want to confirm the location info you sent to your friend Ifunanya yesterday."

Tolu wants to protest, but Ifunanya pleads. She opens her chat history and I can see the location she sent points directly above her house. I ask Ifunanya for hers, and it points back to the place we had just left.

"How is this possible?" Her voice is barely audible.

I have a faint idea. Techno-occultism. These days, highly established occult fraternities use Zoom for meetings. It was only plausible since 'big-men' who made money hardly stayed long in the country. Gone are the days of transforming into crows at night.

This however, is something I haven't seen anywhere before. An app that had seer capabilities? Hijacked location data? Missing girls who all used the app? And the tree was at the center of it all. I wonder where the resident spirit is hiding tonight. It was not in the tree…

"Ifunanya, if you don't mind I'll like to have your phone," I say. "I want to run analysis on the app. I'll be back by noon tomorrow."

I can tell she's relieved to give me the phone.

"One other thing," I add. "Could you let her stay here, Ms Tolu? Just for the night."

"Sure."

"Finally. I'd like you to have this." I brandish a Styrofoam cup from my bag, and from it I pull out three eggs. "Keep them by the door. If this app is what I think it is, then you should be safe tonight."

Ifunanya doesn't complain.

"Are the eggs enchanted?" It's Tolu instead. "Are you sure you're not a scam that'll just sell her phone and run off."

"Ifunanya knows," I say, heading to the street. "After all, she can see the curse on my back."

* * *

It's almost eleven. I call Dele to cash in a favour. He's my tech-guy, and I want him to analyze this app for me. Hive. I'm getting the Eater's vibe all over this, and mixed feelings. After I survived, therapy couldn't save me. For two years I was sent to a Catholic church in Aba. They say Catholics don't pray, I have no comment on that, but their secret rites are something else. Well let's just say I'm now pretty confident of my chances against a wayward spirit.

Even the Eater.

It's funny how the Eater is supposedly the damned soul of a powerful babalaawo from way back. Like colonial masters back. Baba had a medium that told me what had happened to my soul. How the Eater consumes so much so that the princes of the abyss cannot drag it into their pit. I step out of the mikraa, and stretch. God, how do Ibadan drivers fit six people into a tiny cab? Dele stays in Ojoo, and before long I find his flat. Nigerian hip hop music is blaring out of speakers within.

I knock.

"*Who be dat?*" He yells in pidgin.

"Nna."

"*Ahh, witchdoctor himself!*" His burglary-proof door swings outwards. "My guy! *How far now?*"

I ignore him and enter his room. It's stifling, as he uses his room for his business too. The air is thick with mosquito coil smoke.

"Isn't this stuff bad for your health? And your computers too?" He shrugs. I hand him the phone. "I want you to analyze an app for me."

Dele plugs the phone into his myriad of screens and system units. Doesn't he care about radiation and cancer? And noise pollution?

"Interesting," he says after a while.

"Why?"

"*Omo Nna, you go like come back tomorrow oh. This algorithm complex wella.*"

"Complex? It was built by a uni student."

He shrugs, and I have no choice but to go home for the night. It's almost midnight, and Ibadan dwellers aren't too keen on late night services. By the time I get to my place, I'm tired but I can't sleep. The app was the key to finding the missing girls – if they were still alive. Hive. Whoever's running that app is as guilty as those shady babalaawo who deal with kidnappers.

I resolve to track them down tomorrow.

* * *

I'm young again. Filled with hopes and dreams for the future. We just concluded a major test, and coincidentally, the rest of the school is done with exams. Party broadcasts flash over my notifications panel, and I'm considering indulging. Didn't they say 'all work and no play makes Jack a dull boy?'

Besides, I need to drink something to banish this mounting post-failure depression.

One of the gang has a sugar daddy. A chief or something. Says he's throwing a party somewhere in Ibadan. High class, real deal at one super-exclusive club, and he's willing to pay for our entry. I don't really like the idea, but the sugar daddy will probably book us a hotel if we get stuck outside and everything looks good…

The club's name is Hive.

Things are never slow in a club. Everything is fast, loud, and bathed in a flurry of blue and red lights. A dizzying cocktail of loud music and sweaty bodies. A den of lust and liquor. A fancy 'candy' is going around. I pass, because I've had too much to drink and I'm barely dancing away the alcohol in my system. Nobody sees the hooded people at the corners of the room chanting in Yoruba. Nobody notices the drum in the center, or when the masquerade starts dancing with them on the dance floor, blue lights flashing over the dull machetes in its hands. Red liquid rises where the stoned dancers fall. The faces of the masquerade are grinning in twisted bloodlust, life dripping from its dull blades. It punctures the drum, and black smoke follows, hungrily devouring the souls of the fallen, swallowing the lights in its shadows with a trail of fire in its wake.

Then it comes for me.

All I remember afterwards is the burning club. The faint image of a hooded person speaking incantations over my head. Stopping my soul's 'bleeding' as Baba put it. It is still leaking though. The seal isn't entirely complete.

And neither is my soul. Not anymore.

* * *

Mornings are usually so-so, but today I manage to get a breakfast of agege bread, and flag a mikraa to work. I don't look over my leads on that sugar daddy those years ago, or the burnt down 'illegal club' Hive. Or even my mysterious saviour. Today, I'm closer to the Eater than ever before, and in proper tech-guy fashion, I get a call from Dele to pick up Ifunanya's phone.

"*I been work on the app overnight.*" It's evident, but I don't point it out and just listen to what he has to say. By eleven, I'm back in UI, in front of

Tolu's place. I knock.

Ifunanya's head peeks out.

"I got your phone. I need you to take me to the Tech faculty. And we need to talk."

She's ready in five minutes. She seems even more agitated. I notice the eggs are absent, so I ask about it.

"When we woke up this morning, their shells were broken, and the yolks blackened."

Just as I thought. It came for her, but couldn't get to her because she wasn't with the phone. Probably why I survived because I never took the candy. I explain this to her, and she looks even more perplexed.

"Ifunanya you're a seer right? When I analyzed this app, there were empty sections of coding. My contact called it unusual. Incomplete. However, there are actually codes on those empty patches. Codes only seers can see."

"How is this possible?"

"I have no idea. But I think we should head back to the tree first."

However, the tree is no longer there.

Ifunanya immediately wants to apologize but I hold her off. The seer-codes might have hijacked Tolu's location info, and led the girls to the Eater's location instead, but I cannot fully fathom how the tree itself had disappeared. I have a theory I've been working on ever since I left Dele's place, and I'm suddenly itching to try it. I ask Ifunanya to lead me to the app's creators.

The moment I see him, I know he's the one. It doesn't matter that he's in the midst of a crowd. He notices me too, and takes flight. I chase after him, shoving youths with earphones in their ears and textbooks in hand. He's running towards one of the hostels. I'm not fast, and I wholly dislike any form of physical exertion, but I have a few tricks up my sleeve. Almost short of breath, I pull a piece of red string from my back pocket. I'm muttering some Igbo – remember Aba – keeping him in sight at all times. I tie my pinkie and ring finger and bind them, and his legs clap together, throwing him to the dirt. It's a forest shortcut, and we're the only ones here.

"Please!" He's begging all of a sudden.

Ifunanya catches up and he begins screaming for help.

"I want to know about your app."

"Please don't kill me! Please."

"Are you a seer too?" and this time it's Ifunanya.

He doesn't reply.

I bind my thumb and he begins to gag and choke.

"Seer? Oh yes. Please stop. I just needed a good piece of project work done! Chief said he'd help me. Please!"

"Chief who?"

"I don't know his real name. The only person I have had contact with is one of his aides. They helped me with the app!"

"How did you then write the codes? How did you become a seer?" I'm clearly losing my patience.

"A dream!" he cried. "The aide said I should follow the instructions from there."

"And the instructions made you name your app Hive?"

"Yes sir!"

Dream possession! Maybe that's why he can see too.

"And how come the app can run offline?"

"It's a small scale network. It works like transferring files between devices using wireless hotspots."

"All the features?"

"Just chats. The location services also have some glitches."

Convenient.

"Now you will register me on this Hive."

"In the dream. Only UI students!"

I tighten the string on my thumb. "You're an admin, and I don't care."

"Okay! Okay!"

* * *

He lives off-campus. A surprise. He's creating a user ID for me when Ifunanya speaks up again. "I can't find the chat groups for the missing girls."

"It's a small glitch I'm working on," he says anxiously.

"Some people are posting that you're the one deleting them."

"It's a glitch from the system, okay? I'll sort it out."

I take Ifunanya into the corridor. "I think he doesn't really know what a seer is. He might just be a temporary one too. He was probably tricked by whatever was in that tree into writing the codes for the app. He probably doesn't know the difference between the human code and the *other*."

"What about this Chief? The aide?"

"One thing at a time. Ifunanya, I've experienced something like this before. I have a plan, but I'll need your cooperation."

She nods.

I'm in the system. I head back to Tolu's place. I'm inside, and am now trying to send location data over mobile network.

"It's fine," Ifunanya says. "No glitch."

"Perhaps the spirit only hijacks it over the local network. How many girls were missing?"

"About six, last I checked. Including my friends."

"Any new reports today?"

"No. But the groups might have been cleared."

"We'll assume it didn't get any. Seven. Maybe you were meant to be the last one." I think about the candy, and the link to the Eater. The app, and its link to the missing girls. The missing tree.

"I'll be bait tonight."

* * *

It's one a.m. when I get Ifunanya's location info. She's supposed to still be at her friend's place. However I left her phone in my office, so I'm getting the text from Tolu's phone. My client's safety first... I haven't been paid either, so, you know. After leaving Tolu's place, I hang around common rooms, and now, the library, where the local network is strongest.

The location I get however, is not Bini road.

I'm not an idiot. I have a feeling the texts are also monitored, so I had made up an hour of elaborate dirty chatting with Tolu/Ifunanya. Now I wanted to come over under the cover of night and let loose of some steam. Lies, but the Eater doesn't know that. The location is leading me towards a girls' hostel alright, yet the road is quiet, the streetlights are all dead, and even nature is holding its breath. There is nobody on the road this late, and even the security personnel are nowhere close.

The tree lies in the middle of the road. A pool of black liquid is swirling and bubbling from its hollow. I feel my palms freeze.

"*You're not supposed to be here.*" The voice is multilayered, and its sinister chords scratch against the frayed edges of my soul.

"I've been looking for you, Eater of Souls." I sound exactly like I am feeling, and it is not brave.

"*You only escaped once before because you had not exposed yourself to me. Now, however, is a different story!*"

I pull out my wand. The next scenes are high level occult stuff. I will choose to omit this, for your own good.

However, I walk out of it, covered in black goo and blind in one eye. I do not completely vanquish the Eater, mind you. It's been alive for years, devouring humans during that time. However, it flees, and the missing girls are found the next day in their hostels, with mild migraines and amnesia. Hive has crashed too. Turns out seer-codes don't run on all operating systems

Look at that. I cracked a joke.

Ifunanya paid well, and is now my assistant too. I still don't understand how it works, just that the dusty shelf is no longer dusty, hmm. I've set my sights on the Chief, his aides, and this Hive. Spirits cannot take a person unless the person is made aware of their existence... or thrust into their path. This Chief has been the center of Hive, and if what I think is right, then Hive may be an organization of occult groups, and the Eater was just one of many...

I'm tired. Let me end it here.

* * *

A month has passed. Life has become slow again. Ifunanya is at a small window desk, just below the bad paint-job but the now-fixed ceiling. She has stopped social media, but I wonder for how long. A small knock drags me from my thoughts, and I tell whoever it is that the door's unlocked.

It's a boy. He looks like the daily labourers at construction sites.

"Hi. I'm Dominic Uchenna. Call me Mr Nna. The nice lady over there is Ifunanya. My assistant."

"Mr Nna," his Yoruba accent is thick. "My name is Joba."

"Hello Joba. What can I help you with?"

"Last week I went through deliverance. I was possessed by an evil spirit."

Nasty. But this is my job. Believe it or not, people need it. Need me.

"Tell me Joba," and I lean in. "How does that make you feel?"

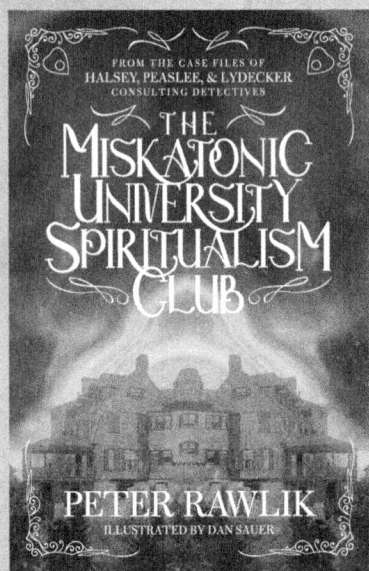

THE MEMORY FUMES

RHYS HUGHES

No sooner had I finished glueing together the two halves of my cloven pet monster, Chives, than an odd bird flew through the window and perched on his head. Chives frowned and rolled his eyes upwards. I noted that his eyes didn't operate at the same speed. Clearly it would take some time for both sides of his frame to integrate themselves properly into a whole. But the adhesive I had used was very powerful and I wasn't too worried about the risk of him coming apart again.

The reason why I wanted one big monster instead of two thinner ones was to avoid paying double monster tax. Personally I think that monsters shouldn't be taxed at all, it's unfair on poor people who have all the love necessary to make a pet monster happy but not enough money to pay the fees. But I don't make the rules. The government does that and nobody is ever sure what criteria it utilizes. I'm not a politician but something much less absurd, an absurdity investigator.

Chives asked me, "Have you ever seen such a bird as this?" He found it slightly troubling, that much was clear. His speech was clumsy and his words were badly joined down the middle, but I understood him without too much difficulty. They would heal.

"No, I haven't," I confessed as I leaned closer to it.

"What species?" Chives cried.

I couldn't answer this question because I wasn't sure. It seemed to be a brand new hybrid. Then it opened its beak and said, "I have a message for Sampietro Mischief. Are you he?"

"Indeed I am!" I shouted in excitement.

The bird nodded and continued:

"You are needed in the great city of Svevo urgently. You will be paid a thousand doubloons for saving it from itself. The people are suffocating on their nostalgia and soon the pollution will spread. This is an ecological catastrophe of the very worst degree!"

Chives made a sour face. "I don't want it standing on my head. I don't like it. Please urge it to flap away…"

"Come now, Chives," I berated him. "It's a beautiful bird and it has an intriguing offer of work for me. How can I be so discourteous? Please sit

quietly and try not to disturb it. I wonder what the exchange rate between doubloons and euros is? Do you know?"

"Ask a pirate," Chives snapped.

When a monster is in a bad mood it is best to ignore him. I said to the bird, "Tell me more about the commission. Who sent you? How are you able to talk to me? This is strange."

The bird replied, "I am a cross between a carrier pigeon and a parrot. For many centuries pigeons have been used to deliver messages, but I'm able to speak them, which is better."

"Ingenious!" I clapped my hands. "A simple but brilliant idea. So you have flown here from Svevo, yes?"

The bird shook its head. "No, I was dispatched by AIR, the Academy for International Respiration, a society devoted to keeping the atmosphere of our planet as breathable as possible. It monitors the cities of the world and takes whatever action it can when the pollution levels get too high. A losing battle, but it *must* be fought."

"I have never heard of this Academy," I confessed.

"That's because it's a secret."

"But where is it based?"

"I can't reveal that. And don't try to follow me when I leave here, for I won't head straight back to base, but take a detour. I have other messages to deliver to many other people."

"I wouldn't dream of it, dear bird, but are your masters aware that I'm very reluctant to leave my home? Svevo is far away. It's tucked up in the furthest northwestern corner of Literary Italy and I prefer to solve cases in my own house. Please tell them this."

The bird flapped its wings but didn't take off.

Chives grimaced. I ignored him.

The bird said, "They know many things about you, Signor Mischief, including your dislike of travel; but they are insistent that you *must* go to Svevo, that the situation can't be dealt with from outside. And they have spent a large amount of money in turning Svevo into an imitation of your house, so you will feel more relaxed."

I recoiled at this and almost tripped over the oboe that I had foolishly left on the floor. "I beg your pardon?"

"Carpenters and theatre workers were employed to create backdrops resembling the interior of your house. These have been erected like walls all around the city of Svevo. And a ceiling has been placed over the top, with a glass skylight that allows the sun to shine through and slide over the ceiling to follow the sun's motion. Thus it mimics an electric light. I believe you will feel at home there."

"But..." I was almost too shocked to speak.

"Yes, Svevo has been enclosed, shut in, merely to keep you happy. It makes the pollution even worse, of course, which is why you must hurry, hurry, hurry to the location, Signor."

And then the bird raised one wing to salute me and took off by leaping into the air and flapping vigorously. It flew once around the room before soaring back out through the window.

Chives wiped the top of his head with a clean cloth. "Come!" I said to him. "We don't have time to waste!"

"You really plan on travelling all the way to Svevo? But the bird never even told you how to collect your fee. I don't trust it at all, nor the shady rascals who sent it," huffed Chives.

"That's an abominable attitude," I pointed out.

"Of course it is. I'm a monster!"

Nonetheless, Chives didn't try to dissuade me further. We packed our luggage and I consulted a book of train timetables, because I had no plans to expend physical energy. Rolling smoothly downhill to Calvino all the way on a bicycle is one thing, but Svevo was even more distant. I wanted to travel efficiently and in comfort.

"There's a locomotive leaving in fifteen minutes!"

Chives hefted my suitcase and lumbered to the front door. The suitcase was full of books, oboes and plaster gnomes. Then it occurred to me that if the pollution was really so bad in Svevo, I ought to protect my lungs. I am human and only have two of them.

"And bring me my diving suit, just in case!"

"The scuba outfit, sir?"

"No, Chives, the deep-sea one. Hurry!"

He winced at my order and shambled off with a titanium scowl. When he eventually returned from the basement with the rubber costume, brass helmet and pump, I noticed they were covered in old soap stains. Had the rogue been using them in the bath without permission? It wasn't the right time to punish him for such violations.

"I have a hunch," I said, "that it's better to be bored and stay alive than to have access to entertainments and be suffocated. So empty the suitcase and replace the contents with this suit."

"What about *my* lungs?" Chives demanded.

"You don't have any. You absorb oxygen directly through your horrid skin. Now we must hurry! The train leaves in five minutes. With luck, it will be late and we will catch it."

And that turned out to be exactly the case...

We raced down the street, arriving at the station just as the locomotive was pulling out, and we managed to jump on board in time. But we didn't have tickets and had to buy them from the furious inspector who stamped up the corridor in his creased uniform.

"This is first class!" he bellowed. "Get out at once!"

"But I'm a first class fellow," I said.

He pointed at a sign that said: NO MONSTERS PERMITTED. And so we were forced to vacate our seats and make our way to third class at the rear of the train, which was a very primitive and filthy carriage full of farmers, chickens, donkeys and vagabonds. Fortunately, they turned out to be fine company, this motley collection of unpretentious passengers. They didn't resent my intrusion and I got on well with most of them and listened with authentic interest to their rustic songs.

The train chugged on in darkness, but we hadn't entered a tunnel. The truth was that this carriage had no windows. Then somebody lit a stub of candle and shadows danced on seats.

I scrutinized my fellow travellers more closely.

A bottle of home-brewed brandy was passed around, not as smooth as the stuff I preferred but good enough.

All of them had memories of some better time, when they were noble and dignified people, not brought low by the cruelties of circumstance. It stirred feelings inside me into a sludge, these reminiscences, and because many of my feelings were opposed to each other the end result was sickly and garish and difficult to keep down.

"Memories," I remarked to myself.

"What about them?" Chives wanted to know.

"Well," I replied, "don't you agree it's a strange thing how memory is so inextricably linked with identity?"

"I don't know what you mean. I have an obvious identity as a monster but I often forget that's what I am."

"You forget that you are a monster? How?"

"The mirror never reminds me. It's not my fault. I peer into it and see a handsome creature gazing back."

Chives was determined to be awkward, but I pretended not to notice. I gave him a brief lecture on the philosophy of Heraclitus, how nothing can remain the same, how everything is movement and change. That is a law of nature and it applies to us. We all assume we are a continuation of the person we were ten years ago but in fact we are strangers. No molecule in my body or yours is the same, all have been replaced, and this holds true for the people we will be in the future.

The Sampietro Mischief who solved the riddles of the missing sky and the Marco Polo clones is *not* the fellow who is sitting on the filthy planks of an enclosed gloomy train carriage.

"It is impossible to cross the same river twice. That's what Heraclitus said," I finished with an air of finality.

Chives considered this statement. "Because the particles of water will have rushed downstream? But what if you run along the riverbank ahead of them and jump into the water again before they reach the sea? I suspect that Heraclitus was simply a lazy man."

I opened my mouth to berate him for his deliberate ignorance, but one of the peasants spoke up. "Your monster is right. Heraclitus was in error. There is no movement anywhere at all."

"I beg your pardon?" I responded in an icy tone.

"Motion is logically impossible."

I hate to be belittled in front of Chives, so I contorted my visage into a grimace and spat the word, "Really?"

"Yes," continued the peasant, as he tugged his tangled beard. "It was explained to me carefully once. A philosopher named Zeno proved that nothing moves, that nothing *can* move."

A painful shudder racked my chest as I sighed.

"With paradoxes," said the peasant.

Chives grinned and encouraged him to give an outline of the infamous antimonies or supposedly insolvable paradoxes of Zeno. I had no patience for this, but I was stuck where I was.

First the peasant discussed the race between Achilles and the tortoise and how the former could never overtake the latter, however fast he ran, provided the tortoise had a head start, because whenever Achilles reached the point where the tortoise had been, the tortoise would have moved on a little and Achilles would have to catch up again, and when he did that, the tortoise would have moved on again.

And so on forever. I often describe myself as a *Zenophobe* because his mathematical puzzles infuriate me. But the peasant wasn't finished yet. It seemed to be his duty to describe a second paradox, the one that insists an apple, or any object, dropped from the hand will never reach the ground, because it first has to pass a point halfway between the hand and ground, and to reach this new point it first has to pass a point halfway between the hand and the halfway point, and so on.

Feeling that a headache was imminent I cried:

"And do you deny that this train is moving? If so, why have you come aboard? Is train travel also an illusion?"

"Yes, but a useful one. When I arrive at my destination I'll know it's a mirage but that doesn't matter. One can enjoy a place that doesn't exist if one happens to be in the right frame of mind." The peasant shrugged and packed a clay pipe with foul tobacco.

"Smoking is banned on public transport in Litalia!"

"Not in the third class carriages. In fact it is encouraged here to mask the other smells. Trust me, I know."

As he ignited and puffed on this disgusting mixture, I became aware of the stench of my environment, the combination of sweat, manure, alcohol and disease, the odour of rotten breath and the pungent foodstuffs trodden into the bare boards by earlier commuters, the stink of decay and poverty, the fogs of unwashed socks and shirts.

"Open a window, Chives!" I commanded.

"But there are none," he said.

I had forgotten that simple fact. Yet I wasn't going to be defeated by a surly pet monster. "Very well. Then you must climb onto the roof and sit there with the air pump while I wear the diving suit. You will pump fresh air to me while I recline in this mess."

Chives turned pale and his teeth chattered.

"Surely you don't intend to send this poor fellow outside?" protested the peasant, and most of the other passengers grumbled in solidarity with his outrage. "What if he falls off?"

"According to Zeno, he'll never hit the ground."

And I felt the warmth of malicious satisfaction flow through my veins as the peasant tugged his beard harder in dismay, his pipe glowing as he puffed on it with renewed vigour.

"But he has a crack down the middle."

"True enough, but I have rejoined him skilfully. So there is no danger of him coming apart. He is whole."

"But it's impossible to join two halves of a broken monster. First one half would have to pass through a point halfway between it and the other half, but to reach that point it must..."

"To hell with your Zeno!" I shouted.

Chives said in a quiet voice, "If Zeno went to hell he would first have to pass through a point halfway—"

"Enough! Is this an open rebellion against me?" I stood and bitterness flooded my heart as I pulled Chives to his feet. "Have you forgotten who gave you life? Now climb onto the roof with the pump and never question my instructions again, do you hear?"

He perceived my anger was at a serious level and so he meekly obeyed my rather callous command, climbing a rusty service ladder bolted to one wall of the carriage and pushing himself through a trapdoor in the ceiling. I had a sudden vision of sky and rushing clouds before Chives reappeared in the gap, his arms thrust down to receive the pump. Standing on my toes and stretching, I passed it up to him.

The breathing tube now connected the fresh exterior with the stale and nauseating interior. I climbed into the bulky suit, lowered the helmet over my head and sealed the little porthole.

"Start pumping!" I shouted, but my voice was muffled, trapped within the confines of the brass helmet. Yet Chives used his own initiative and I felt the rush of clean air over my face.

I sucked sweet oxygen into my tortured lungs.

The peasant who had given me the lesson on Zeno scowled and puffed harder on his pipe. I was no longer acceptable company to him or to his companions. I had become arrogant, pompous, swollen with hubris. But I felt that I was a pioneer, the discoverer of a new travelling experience, for the benefit of everyone, not just myself.

Encased in my diving suit because of the stench, smeared with the filth of the farm beasts that strolled around with perfect liberty, I used my nose alone to appreciate our progress, for my eyes and ears had been deprived of the usual stimulus that accompanies a voyage. With Chives perched on the roof and working the pump, the flavour of the air he gave me changed as we rattled through the unseen countryside, allowing me to experience a purely *olfactory* journey through Litalia.

Haystacks, orchards, meadows full of flowers. And darker smells too, mercury mines, isolated factories, graves.

And so I hurried through the scents of our land without moving from where I reclined. And I imagined that I could remain like this all the way to Svevo, but in fact the train stopped outside that city and the passengers began lining up to wait for the doors to be unlocked. Chives came down the ladder and I opened my faceplate and blinked at him in the miserable candlelight. Then sunbeams poured in...

A guard had opened the main door from the outside. The peasants and animals jumped down without saying farewell to me. I had become some sort of villain to them. When they had all gone and the carriage was bare, I stood on the threshold and frowned at the guard. He was the fellow who had inspected our tickets in first class.

"Why are we not proceeding to the main station?"

"Too dangerous, that's why," he said.

"So this is the final stop?"

He nodded. "Yes, and if you don't disembark here I'll charge you and your monster for the journey back."

Then he sauntered off down the line, hands in his uniform pockets. In my bulky suit it was difficult leaping down onto the gravel. Chives came directly after me and we staggered across the tracks and up a slope. At the summit we had our first view of Svevo.

"It's like the façade of your house!" cried Chives.

And so it was. I was appalled and comforted at the same time. It was a clever illusion and I felt I was standing just across the street from my own home, that half a dozen steps would take me to the front door. But in fact there was only a painted board of immense size and the windows, shutters and balconies, and even the roof garden, were two dimensional, flat and useless. Beyond this frontage was the real city with all its houses pouring smoke and blotting out health and hope.

"How far away do you think it is?" I asked Chives.

"Five kilometers," he estimated.

"If I take off my suit and pack it in the suitcase, you will have to carry it to our destination. It seems to me that you might as well give me a ride on your shoulders, for the weight will be the same. I have a feeling I will be needing the suit when I pass through the front door and into the living city, if what the bird said was all true."

"Does the sight of your house make you feel less homesick?" Chives asked as he cantered towards it like a strong goat, sagging only a little on his three legs under the bulk of my shell and me. I was about to answer in the affirmative, but then I realized that in fact the replica made me just a little anxious, as if my real abode had been stolen and relocated to a part of Literary Italy I had never explored.

It took an hour to reach the front door and when Chives lowered me to the ground I was reluctant to pass through. I felt like an uninvited guest, a salesman, and yet this façade had been erected for my benefit. My arcane employers, the mysterious people who controlled AIR, had gone to great trouble on my behalf. Fear was clearly unnecessary, so I ordered Chives to remain outside while I entered alone.

The gigantic front door was shut, the handle out of reach high above, but there was a cat flap cut into it at ground level, and I was able to pass through this without scraping my suit too much. What I saw on the other side bewildered me. Ten million fumes.

And each fume was different to the others, a slightly coarser texture or a slightly smoother one, a different way of spiralling or clotting; and the colours were garish, unreal, confusing.

I blinked at a landscape of chimneys, tall thin cylinders poking out of the roofs of houses, factories, public buildings. And every chimney was constructed in the form of a cigarette. It was a heaven or hell for smokers, possibly both combined, a nest of cigarettes like the spines of a nicotine god, a hedgehog of toxins. And yet what created these fumes was a habit that nearly all of us indulge every day.

Memories. That's what was being burned here in hearths and grates and furnaces and braziers. The unwanted memories of yesterday and the days before yesterday. The people of Svevo were burning their memories so that they could start afresh every morning, with a new identity, for the link between memory and identity is such that without the former there is no latter, in other words if you no longer remember who you are, you are finally free to become somebody else.

I knew all this because my faceplate was open and I was compelled to inhale many of those fumes. I absorbed the poisons they contained and I felt nostalgic, sentimental, angry; but none of these emotions were mine. I was breathing second hand memories.

Coughing and spluttering, I passed back through the cat flap, my brain overwhelmed with discarded images and feelings, with incidents from the recent past, from the yesterdays of the million inhabitants of Svevo, every cup of coffee brewed, every meal digested, every accident, every kiss and argument, every minor detail that had seemed significant at the time to its owner but had been incinerated today.

"Your face has turned green," commented Chives.

"I nearly choked in there," I said.

"It was meant as a compliment. No matter."

When I had recovered my breath I planned my next move. "Somehow I must persuade them to stop burning their memories. But how? I can use persuasion, of course, or trickery. I might even try to extinguish each fire myself, but that would be a huge task."

Chives had a suggestion. "Fires need oxygen to burn. Svevo is encased by a ceiling and walls, but there are open windows high above. Shut those and the oxygen supply will be cut off."

He pointed up the façade to the highest floor of the fake house. He had noticed what I had not, that the windows on that level were open. Air was rushing through them from outside, feeding the fires within but making it impossible for the smokes to leave against such a forceful flow. So Svevo was consuming itself at an increasingly faster rate. Time was indeed short and I had to make a decision quickly.

"My mission is to shut those windows, but I won't be able to climb up so far in this heavy suit. It's exhausting. I need you to carry me, Chives. I also need you to operate the air pump."

"I can't be inside and outside at the same time," he pointed out, but he expressed his worry in the normal monster way by rotating his horns. He feared I had a solution to that problem.

"Yes, you can," I said quietly.

"You can't split me back into two! The glue you used was too strong for that. You'll never break the seal..."

"Poor Chives. Don't you remember what the peasant on the train said? Zeno proved that you were never really joined together in the first place, so it will be easy to break you apart."

"All the same, I prefer to remain in one piece."

"Don't be selfish, Chives!"

He attempted to convince me that a single monster is more useful than two smaller ones, but it was futile. At last he submitted to my will and I broke off a branch from one of the dead trees on the ugly landscape and used it as a lever to split him in half.

It was easier than I had anticipated, thanks to Zeno.

"One day I must visit the birthplace of that philosopher on pilgrimage, to express my admiration," I said.

But the left half of Chives replied, "You would never be able to reach that destination, for you would first have to pass a point halfway between the start of your journey and his birthplace, and to reach *that* point... Do I need to say more?" And no, he didn't.

The right half of Chives had the extra leg, which means it was suitable as a mount. The left half would stay here and work the pump. It might be deemed unwise to split a monster in half for the reason that the good and bad in its character might cluster on different sides, that one semi-Chives might be an angel and the other a devil.

The truth of the matter is that monster morality isn't really compatible with that of humans. I have never been able to distinguish between those times when Chives has been genuinely helpful and when he has been only devious. In fact I suspect he can't either.

I closed my faceplate, waved at the one-legged half to begin pumping and pushed the bipedal half through the cat flap. I followed closely and on the other side I mounted and urged him forward at a trot. The rubber hose that was my lifeline reeled out behind...

"Tally ho!" I shouted in my muffled voice.

We entered the dense fogs, the multicoloured swirling clots of vapour, the memory fumes, proceeding down unseen streets, almost bumping into spluttering people who lurched along with cloths pressed ineffectually to their mouths. A few of the richer merchants wore gas masks. And yet no one seemed inclined to stop burning their memories. They were addicted to the feeling of renewal, to pure hope.

That was the root of the problem and I learned this merely by keeping my eyes open, though what I saw was very blurred and hampered by the thick glass of my porthole as well as the pollution. If men or women are dissatisfied with life, with their present identities, and wish to change, a simple burning of memories will set them free, put them at liberty to start again from scratch, like newly born babies.

That feeling of unlimited potential must be exhilarating in the extreme. But the moment a movement is made, irrespective of Zeno, any decision is taken, the future possibilities are lessened, a clear path starts to emerge, and this new direction is no less constrained than the direction of the life that was considered unsatisfactory before.

The temptation must be overwhelming to consign the memories of *this* life to the flames also, to begin afresh a third time, and then a fourth, fifth and sixth, and so on. The habit is soon unbreakable. Every day memories will be burned, turned into spiralling smoke.

How had the people of Svevo discovered that memories could be used as fuel, that they were highly combustible?

That detail was a mystery and would always remain so.

I suppose it was an accident, the first time: a devastated lover, a failed businessman, a victim of an assault. Imprisoned by the identities they had and desperate to escape, to be renewed.

Somehow, perhaps while lighting a cigarette or a lamp or a stove, and leaning forward at a specific angle, a memory had fallen out of the brain, slid like an eel out of the soul, and fallen into the flame. Whoosh! Like a pinch of gunpowder it had instantly flared up, vanished. Once the skill to burn memories was acquired by one citizen it would soon spread. That is always the way. All pain gone in a puff...

One thing I noticed as I rode Chives through the streets was how those citizens now stopped and stared at us.

Many of them pointed at us and even laughed.

I was dismayed by this reaction. What was so funny about a man in a filthy diving suit riding half a monster?

In the central plaza of Svevo we found a public fire.

People were using pitchforks to cast raw memories into the inferno, a pile of squirming images, of love and tears and domestic incidents, like a gathering of heretics in a barbaric century, and a few of these condemned memories protested, pleaded, screamed.

They didn't want to be destroyed. They wanted to exist.

But they were shown no mercy.

"Some memories are unwilling to go," I told Chives, but I'm not sure he heard or understood me. And his sympathies were generally peculiar, so it was better not to expect him to share my horror at this scene. Yet I was forced to conclude that it was none of my business if the citizens of Svevo wanted to deny the past in this way.

A city of chimneys, each chimney like a giant cigarette. And citizens like smokers, anxious, pinched, greedy.

I remembered that I was being paid for a task.

The people with the pitchforks noticed me, began laughing. Then they stopped adding more memories to the fire. The laughter became their new occupation, requiring all their energy.

We galloped past, Chives and I, the laughter following us down streets on the other side of the plaza. We were searching for the stairs that would take us to a higher level, to the floor where the windows were open, but it eluded us, that flight of simple steps.

"I was led to believe the city would mimic the layout of my house, but that was an exaggeration," I grumbled.

"There are no stairs anywhere," confirmed Chives.

"Bystanders are laughing at us!"

"But we are no joke. We are a serious pair."

"Yes, it's incomprehensible."

I wondered if it was an old custom of the citizens of Svevo, this weird hilarity at the arrival of strangers, or whether there was something special about our appearance, an oddity of presentation. It was pointless stopping to engage anyone in conversation; we had no spare time for that. Locating the stairway and closing the windows was our priority, to make the fumes thicker for a short while before starving of oxygen the fires that produced them. I urged Chives to run faster.

We explored every street, every alleyway.

And the smokes grew thinner as we proceeded, as if the fires burning under the memories were dying of their own accord, or as if no one was burning the latest batch of memories.

But there was no obvious reason for this improvement.

I decided it was a coincidence.

Then we were suddenly at the cat flap in the front door, back where we had started. We had managed to tour the entire city without finding a stair or any other method of accessing the higher floors. And this was hardly a surprise when I pondered matters more carefully, for the façade was false, not the front of a real house at all, merely a painted scene, no more solid than a theatre backdrop. So I dismounted.

"We are wasting our effort here, Chives. Let's go outside." And I went through the cat flap into the sunset.

I peeled off the diving suit with relief.

Chives followed me and he greeted his other half, who was exhausted from working the pump handle with his single arm. I had no glue with me to attach both parts together, so I pushed them both into the diving suit as a temporary measure, allowing the rubber to compress them into a single being, though the extra leg didn't fit properly and had to be curled up in a painful position. Chives grimaced.

"A failure, this case," he said rather sadly.

"Some of them are," I agreed.

We began walking towards the railway.

A bird flapped down from the sky and landed on Chives, on the top of his head, just as it had done before.

It was the parrot crossed with a pigeon and it seemed pleased. But this couldn't be true, surely? We hadn't succeeded in our quest and I pointed this out as I pulled off my helmet.

"On the contrary," it said, "you were marvellous."

"I don't comprehend you..."

"You did everything precisely the way the Academy for International Respiration hoped you would. The thousand-doubloon reward is already waiting for you back at your house in Buzzati. You will have to hunt for it in every room. That will be fun."

"It doesn't seem right to take the money. I achieved nothing. I merely rode in a big circle through Svevo, the smokes clearing as I proceeded. I extinguished no fires and persuaded nobody to hold onto their memories instead of burning them. I failed."

The bird flapped its wings and said, "You are wrong, wrong, Signor! When they saw you galloping down the street on your monster, dressed in that diving suit, smeared with the filth of a third class carriage, trailing a rubber tube behind you, they suddenly knew that some memories must be preserved. No one who witnessed your passing today would want to erase from his mind the image of that."

I was uncertain whether to be flattered or insulted.

"Because I was so unusual and amusing, original and odd, they will want to preserve my image as long as possible. That's why they stopped burning their memories?" I cried.

"Yes, yes! That's right."

"And the pollution problem?" asked Chives.

"Will be cured," said the bird.

"So I was a puppet of your masters?" I snarled. "They anticipated how ridiculous I would look entering Svevo like that; and they even knew the exact circumstances of my journey, that I would take the diving suit with me to protect my lungs and that on the train I would be forced to sit in the third class carriage with the animals..."

"Yes," confirmed the bird, "and also that you would ride only half of your monster while leaving the other half outside. They are clever beings, the management of AIR, clever indeed."

I felt bitterness rise up inside me. I growled.

"I have been manipulated!"

"So now you know what it feels like," said Chives.

It was pointless complaining.

I fell silent because I had nothing more to say.

The bird stretched its wings.

And it flapped away. I watched it dwindle in the sky.

"A pawn, I was a pawn!"

Chives grinned. "And I was a knight, the piece that looks like a horse, and we conducted a knight's tour of the city and checkmated ourselves in the final move!" He had extended my metaphor in a way I didn't like and into obscurity. I didn't answer him.

Then carpenters appeared and began dismantling the façade of Svevo. They must have been hiding in the landscape until now. They leaned tall ladders against the false front and climbed up them and began sawing and hammering and discarding pieces. I felt I was witnessing the destruction of my real home and I turned away.

We trudged to the train tracks. When a locomotive finally arrived, we boarded it and sat in third class again, but I remained silent until we were back in Buzzati. I was impatient to be home, worried it wouldn't be there, or that inside would be a million tiny citizens. But nothing had changed. I took refuge in my softest, deepest chair and played melodies on the oboe until midnight. Then I went to bed.

I found it impossible to fall asleep. Who were the mysterious members of AIR? I desperately needed to know. I rose and went to the room where I keep the books I haven't yet read.

I selected volumes at random. At last, in the only surviving copy of a scroll that had somehow escaped the fire when the Library of Alexandria burned down, I found an illustration that solved the riddle. It depicted the philosopher Zeno at the head of a conference table, with Achilles seated at the other end, a glass of mineral water raised to his lips. And between them, on the remaining chairs, gesticulating and perhaps contributing the best ideas, a dozen or more tortoises.

COLD CASES

THE PSYCHIC DETECTIVE: THE FRANCIS ST CLARE & FREDERICA MASTERS STORIES BY R. CHETWYND-HAYES

DAVE BRZESKI

'Someone is Dead' The Elemental and Other Stories (1974)
'The Wailing Waif of Battersea' The Night Ghouls (1975)
'The Headless Footman of Hadleigh' Tales of Fear and Fantasy (1977)
'The Gibbering Ghoul of Gomershal' The Fantastic World of Kamtellar (1980)
'The Astral Invasion' Tales from the Dark Lands (1984)
'The Phantom Axeman of Carleton Grange' Tales from the Haunted House (1986)
'The Cringing Couple of Clavering' Tales from the Hidden World (1988)
The Psychic Detective (1993)
'The Fundamental Elemental' Looking for Something to Suck: The Vampire Stories of R. Chetwynd-Hayes (1998)

When I recently edited *Shadmocks & Shivers: New Tales Inspired by the Stories of R. Chetwynd-Hayes*, there was one creation of R. Chetwynd-Hayes that we were not given permission to use, at least not in a new story. This was his psychic detective team, Francis St Clare and his assistant Frederica Masters. I elected to go with one of the original stories as the reprint by the original author we wanted to feature.

As can be seen on the cover of *The Psychic Detective* – the only novel featuring St Clare & Fred – it was optioned for a Hammer film. Sadly that never came to fruit. The estate of R. Chetwynd-Hayes, however, continue to hold out hope, over two decades later, that there might still be a chance of

R.CHETWYND-HAYES
The Psychic Detective
NOW TO BE A HAMMER FILM

selling those characters to a film, or TV company, and they'd "rather the waters weren't muddied by new stories at this time." Personally, I think that ship has likely sailed by now, and if I'm honest, I'd almost rather it didn't happen, as I suspect the concept and characters would be so changed by updating that we'd barely recognise them. Although, if they made it as a TV series now, in the Netflix vein, and kept it in period, The Psychic Detective would make a superb pilot feature.

The first story, '*Someone is Dead*' introduces us to the characters in a way that immediately led me to envision them as an ITC TV show of the early 70s. Indeed the description of the two primary characters could hardly be more early 70s British TV. I quote...

He was a tall, lean young man with a pale face and the smile of one who is hiding his natural shyness under a mask of easy self-confidence. The girl by his side was extremely pretty: ash blonde hair, white skin, and wearing an expression of cynical amusement as though her blue eyes had seen more than her years warranted. In contrast to her companion's neat black suit, she wore a colourful costume that bordered on the bizarre. The mauve blouse had a dangerous split down the centre that revealed the valley between her breasts; there was a corresponding parting at the rear which offered the masculine eye a tantalizing glimpse of a white, smooth back. The black miniskirt was the stunted offspring of a broad belt and her splendid, nylon-clad legs riveted every man's attention and raised a storm of feminine envy.

Yes, there's an element of sexism there. One has to remember that this was very much influenced by the likes of Emma Peel, Tara King & Sharron Macready. Strong adventurous women were the new in thing – but they still had to be sexy, they had to appeal to the male viewers.

The story itself is not quite what one might expect. Rather than a classic ghost story, it's a 'timeslip' story, which involves people from the past using occult means to attempt to inhabit the bodies of people in the present.

I read a comment on an online forum, from author, and major Chetwynd-Hayes fan, John Llewellyn Probert, in which he gave the opinion that these stories might not be served all that well by being collected together in a single volume, as they had too many similarities. I could see what he meant when I read the opening paragraphs of '*The Wailing Waif of Battersea*'. His description of Fred is almost word for word the same as in the previous story, apart from a variation in the colours of her outfit – and the same for St Clare a few pages later. Another repeated element is the fact that "nothing takes place before 9pm."

Apart from those similarities, the story itself is quite different, albeit the haunting St Clare is called in to investigate does involve human occultist shenanigans. We also see more of St Clare's pseudo-scientific methodology here, as he not only employs the standard automatically triggered cameras, seen in so many haunted house stories, but he has a large gadget of his own invention, which he refers to as a *phantocage*. Constructed from metal tubing, it's rather reminiscent of Carnacki's *electric pentacle* – albeit it's designed to trap an entity, rather than form a protective barrier for the users to safely ensconce themselves within.

I found this to be rather superior to the first-published tale. More of St Clare and Fred's quirky relationship comes to the fore here too.

It had become quite obvious by this point that reading the stories in published order wasn't going to present them chronologically. Chetwynd-Hayes loved to employ the classic Sherlock Holmes trope of referring to past cases, which were never written up. He also mentions a number of cases which were actually published. This story mentions '*The Headless Footman of Hadleigh*' and '*The Phantom Axeman of Carleton Grange*', which weren't published until 1977 and 1985 respectively! Conversely, this story is referenced in '*Someone is Dead*', despite its being published a year later. At this point, I was considering attempting to produce a chronology, once I'd read them all...

While they may be in more or less reverse order of chronology, these tales do seem to get better as the series progresses. In '*The Headless Footman of Hadleigh*', the haunting is caused by something under the house – something so old, that it may have been there since before Stonehenge was erected. St Clare refers to it as an *ele-monster*, but there are elements of an almost Lovecraftian nature. The banter between the two leads continues in a manner which might be construed as sexist, but actually serves to underline just how much of a front St Clare's seeming arrogance actually is. Carnacki used to freely admit to being terrified; Francis St Clare tries his best to keep a lid on it.

I'd read '*The Gibbering Ghoul of Gomershal*' quite a few times in recent memory, since that happens to be the St Clare story I elected to reprint in *Shadmocks & Shivers* (chosen, I can't deny, because it was the only one that wasn't too long for the anthology that hadn't been reprinted fairly recently).

Any hope of putting together an accurate chronology of these stories hit a bit of a brick wall here, as the daughter of the man who requests St Clare's help with the haunting claims to have read the published chronicles of '*The Wailing Waif of Battersea*' and '*The Headless Footman of Hadleigh*', which is interesting since this story itself is referenced as a previous case in '*The Headless Footman of Hadleigh*'!

The phantocage is employed again, and Fred mentions the fact that its only previous use was to trap the Hadelmonster in 'The Wailing Waif of Battersea'. In that story, however, St Clare states that it's never been known to fail, but Fred expresses some doubt that it'll hold a Hadelmonster – which suggests it certainly had been used in other cases. Despite the inter-story contradictions, and yet another colour variant on exactly the same description of Fred's outfit, this is another enjoyable tale – with the intriguing variant of a suggestion that they might not have completely eradicated the problem. It's also notable that the lady of the house has cause to sternly admonish St Clare and Fred over their endless bickering. They explain that it's very much a release valve for the pressure they encounter in their work.

'The Astral Invasion' takes a different tack to the other stories. Here, St Clare and Fred are manipulated into the position where Fred's not inconsiderable psychic power can be subverted in an attempt at bringing about an invasion of nasties from the other side. We get more hints of the true regard St Clare holds for Fred in this one, despite the disparaging remarks and bickering. These complexities of their relationship really are quite interesting.

It's logical to assume that writers become better at their chosen profession over time. I enjoyed all the St Clare stories up to this point, but something happened between 1984, when 'The Astral Invasion' was published and 1986, when Chetwynd-Hayes gave us 'The Phantom Axeman of Carleton Grange'. Where the others are very good, fun stories, which perhaps tend to suffer from a little too much repetition, this one is so much better! He put much more work into exploring exactly what these two flippant young people were actually capable of, and lays very interesting clues to their origins and nature. Fred even gets to wear a totally different, rather less revealing outfit!

Having progressed from the variant colour-ways of Fred's original outfit, Chetwynd-Hayes does rather fall into old habits with the next story, 'The Cringing Couple of Clavering' in that she's introduced wearing a variant coloured version of the same outfit as in the previous story. While some might find this repetition a touch lazy, it does rather answer any accusations of prurience that the modern reader might claim – in truth, it seems this aspect was of little real interest to Chetwynd-Hayes, hence the constant recycling of descriptions.

As with the previous one, this story is again rather more substantial than the earlier outings. It features a major crossover within the author's canon, in that St Clare and Fred find themselves investigating a case at Clavering Grange. Fans of the works of Chetwynd-Hayes will be very familiar with the place, as he set rather a lot of stories in that great manor house that is built on *tainted ground*. The hauntings here go way beyond anything the couple have faced before, involving a group of immortals who are simply trying to get themselves in the best possible condition to handle some sort of important task. I haven't read enough of Chetwynd-Hayes' work as yet to know if this concept is explored further in the many other stories of Clavering Grange, but it seems a distinct possibility.

Which brings us to the only novel length outing for St Clare and Fred, *The Psychic Detective*. There's no mention of previous cases in this one for the simple reason that it's a prequel. It covers the first meeting of Francis St Clare and Frederica Masters. We finally find out more about her origins and family. Having rescued an eighteen year old Fred from a potentially hazardous situation with a much older man, St Clare soon realises that Fred is exceptionally gifted, and takes steps to recruit her to his cause – that of becoming *The World's Only Practising Psychic Detective*. One of their first tasks is to clean his house of a number of unwanted psychic manifestations. In the training St Clare puts Fred through to effect this task, she learns to send her astral form through various levels of reality, or dimensions. As things become far, far more hazardous to not only Fred, but all of reality, the story charges full-tilt into a stunning metaphysical adventure. Both Fred and St Clare have to travel to the Hidden lands and the Ultimate Mind, where Francis St Clare meets The Author.

Chetwynd-Hayes' storytelling takes on an almost Lewis Carrollesque feel as logic becomes somewhat warped. Fans of the author will also be pleased to find that one of his other characters, the redoubtable Madame Orloff, makes a welcome guest appearance here. The early stories were good fun, and the later two more substantial, but this is where Chetwynd-Hayes really hits his stride. I think this might be my favourite book of his that I've read thus far.

Having said that, despite this being a prequel to all the previous shorter stories, I don't really recommend reading this first. If you want to read only one of the stories, then this novel should absolutely be the one, but if you want to read them all, I think published order is best, as revelations about the characters obviously originally came in that order.

And that would have been the end of this series review, except five years later, in 1998, we get one final St Clare and Fred story in a collection of Chetwynd-Hayes' vampire stories. '*The Fundamental Elemental*' is odd. It's the shortest story in the canon by quite a margin. It's also going to be fairly confusing to anyone who hasn't read *The Psychic Detective*, as it opens with St Clare and Fred back in the Hidden Lands, searching for Rowena, a friend of Fred's who is being pursued by a vampire. Frankly, it's a little confusing even if you have read *The Psychic Detective*. The vampire is resident in Fred's subconscious and the method of his defeat is an intriguing variant on the idea of all creation having started with a thought, that became an idea, that begat a theory etc. We discover something about The Hidden Lands that poses far more questions than answers.

Sadly, if Chetwynd-Hayes ever explored the Hidden Lands further, and you can be sure he did, it didn't involve St Clare and Fred.

I still hold out hope, one day, of convincing the Chetwynd-Hayes estate to allow me to license a collection of new St Clare and Frederica Masters stories.

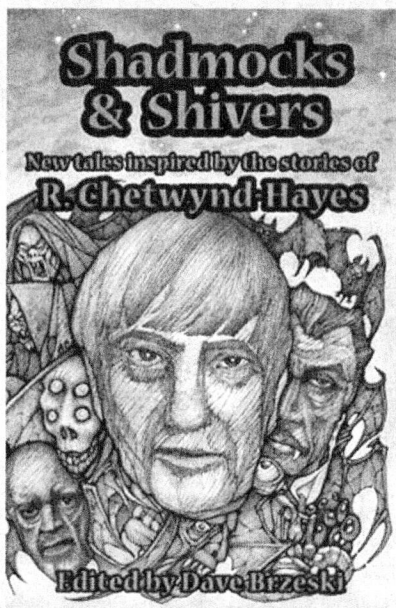

THE VOICE ON THE MOOR

MELANIE ATHERTON ALLEN

For those who, like me, seek the charms of solitude, the moors around Blankford are an as yet matchless delight. The local folk are rarely encountered upon the moor, and so there is a silence – a hush – to the place that fills the soul, almost as if the silence itself were full of voices, unheard yet, somehow, present...

— M. Atherton, *Moorland Tramps, 1908*

I'd meant, this time, to visit my aunt. I had everything planned to a nicety. I would catch the 2:15 from Paddington, which would deposit me at the small town of Blankford at 5:30. Ample time, so I thought, to catch a six o'clock train to Yandover Parva. And at Yandover Parva (I shall trustingly confide to my readers) dwells my aunt.

It was the matter of catching the six o'clock train that undid me. I arrived at Blankford at 5:30 as advertised. Then, instead of waiting sensibly on the platform, I decided that my legs could use a stretch, and that the nearby streets looked romantic and appealing in the dying afternoon light. So I strolled. I have admitted it – abjectly and repeatedly – to my aunt. Now, I admit it to you. I strolled. And, in strolling, I became, in some manner that I find it impossible to describe or to re-construct, hopelessly lost.

"Oh, but Pendleton!" cries the Well-Travelled Reader, "this is all surely quite impossible. Not only is Blankford (where I recently spent a fortnight) laid out on a grid system – rendering getting lost difficult to say the least – but it also could never look romantic and appealing, no matter what time of day it was. Beastly hideous maze of villas, only it ain't a maze."

And to you, reader, I say: wait. It turns out, probably, all to be due to the inscrutable workings of Fate. The romantic atmosphere, the getting lost in a perfectly uniform grid system – all of this – Fate, waiting with a sandbag.

Because, once I realised that I'd missed my train, and knowing that there simply wouldn't be another until the next day, I walked directly into Simon Wake.

* * *

A word of explanation, for readers who are not familiar with these narratives. Simon Wake is an occult investigator. I know him from my club, the Anacreon, where he often dispelled the tedium of the pre-dinner hour by telling us (myself and Vincent Anderson) tales of his supernatural adventures. We'd sit by the fire in the Anacreon smoking room, listening to his wildly improbable accounts, feeling all the warmer and more contented the wilder and more perilous his stories became. Anderson claims he always believed these tales – and what I know of Anderson's simple nature suggests that this is probably true. I merely thought they were jolly good yarns. More fool me. Turns out they were true, after all.

In recent years, instead of telling us about his adventures, the beastly fellow is now just as likely to get us involved in them. Sometimes he sends for us. At other times (as now), we simply find that, through an improbable series of coincidences, Wake is amongst us once more.

Oh, and I suppose I should mention about the whole disappearing business. Wake *vanishes*. This is because he is – though I hate to say it, for it sounds, to my way of thinking, too utterly *silly* – in thrall to Fairy, ever since an investigation of his went rather disastrously wrong. Anyway, now Wake's permanent address is Elsewhere, and he disappears there every night at midnight.

Right. I believe that's all you need to know before we begin. I think we may now get on.

* * *

"Wake," I growled. "I ought to have guessed."

"My dear fellow," Wake said. "It almost sounds as if you weren't happy to see me." Simon Wake, I noted in a brief, horrified survey, was looking stranger than ever. There was an almost transparent pallor to his skin, which seemed to glow against the dark background of his wild black hair. He looked like something out of a Burne-Jones painting. Not nice, some conventional part of me whispered, meeting a thing like Wake in a drably respectable place such as Blankford.

"Where's—" I got that far before I was forestalled by a voice from behind us.

"Wake, by gad! And is that Pendleton? Well, well!" A villa door was thrown open. Anderson's gigantic frame filled the doorway. He was halfway out of trout-fishing kit. "Do come in!"

We came in.

"And what," I asked coldly, "are you doing here, Anderson? And don't say fishing. There isn't a decent stretch of stream for miles."

"Well," said Anderson, apologetically, "fishing."

"But—"

"I know. Not for miles. Not, in fact, for exactly three and a half miles. Three and a half miles from here, however, there is quite a good trout stream. You get the right to fish in a particular piece of it by leasing this particular villa. I've taken the place for a month. Just got home from rather a good day." He glanced down at himself. "You chaps sit somewhere. I'll change."

Soon, respectably clad and cheerful, he was with us once more, and we were all sitting in a slightly gloomy little parlour with glasses in our hands and pipes going.

"Well, now," said Anderson, puffing luxuriantly. "Out with it, Wake. What brings you here? Anything – odd?"

Wake leaned back. He looked tired and careworn. "You're right, of course. I am investigating. At the urgent request of some of the town worthies."

"I cannot imagine," I said, glancing about the painfully new room, "anywhere less likely to be haunted than here."

"Don't be narrow, dear boy," said Wake. "It very well could be haunted. After all, people *do* live here. Some of them even quite *like* living here. And I don't see that your beastly little flat is much of an improvement, anyway. As it happens, however, it isn't here – in the town – but outside of it – on the moors surrounding the place – where the problem lies. I hope," Wake looked at Anderson, "that you haven't been cutting through the moor to get to your bit of stream?"

"As it happens I haven't," said Anderson. "The people who let me this place made me promise not to. Admitted it was a shorter way, but said I was likely to get bogged down and maybe drown. So I said I'd keep to the road, and that's what I've done."

"They told you it was boggy, did they?"

"Yes. I've had a look, though, and it doesn't look too bad. Still – a promise is a promise."

"It *is* boggy, though," said Wake. "In places. But – well, let me give you a sketch of the landscape. The town of Blankford is in the northern part of the moor. There are three distinct and separate bogs, and, though each is almost improbably deadly if once entered, they are all very easily avoidable. One bog is in the western part of the moor, one is in the southern part, and one in the eastern part. No path passes closer than a half of a mile to any of them, and most give 'em a wider berth than that. Also, each bog is rigorously signposted. In fact, the moor ought to be safe. All one has to do is keep to the paths.

"And yet, recently three men drowned over the course of a single fortnight. That was a bit thick, three deaths right in a row like that, so some local worthies got together to talk it over, and eventually they sent for me. Odd, that. I'm not, I think, famous. They must have asked around quite a lot before they came to my name. Anyway – they sent me a letter." Wake paused, and his eyes widened as if at some remembered shock. "It reached me, too. Not – not at my London address, either. I wonder," he continued, and his voice was strained, "if you'd like to see the envelope? It is – a bit of a curiosity." And he removed it from an inner pocket and handed it to me.

I have kept that envelope. I have it by me now as I write. It is an improbably disreputable-looking thing. It is addressed to Wake's almost totally abandoned London home, but things appear to have happened to it to divert its path.

The postage is not cancelled, but half the stamp is missing. There are marks of inky fingers all over the envelope. A horse clearly stepped on it, and may have carried it some considerable distance before kicking it off of its hoof. A corner is missing – looks as if something had it by the corner and dragged it, through the mud that is still very much in evidence. There is a smell of drains hovering round it. A thin powder of fine shiny dust stubbornly adheres to it. If you could reconstruct the route this envelope took by these trace contacts, you would know how to enter Fairyland. A beguiling thought on a grey day, I admit.

But I digress. I shall replace the envelope in its drawer; then, I shall get on.

We handled the envelope for a moment, looking at the various evidences of travel I have mentioned above. Then, into the silence, Anderson asked, "why'd they want *you*?"

"Because the moor is – well, haunted, is what they said. It isn't haunted, I think. Merely – inhabited." Wake's brooding gaze fell upon the reproduction of The Soul's Awakening above the fireplace, and he smiled secretly up at it, as if he and it shared a private joke. Smiling like that, he looked rather like a faun. I shivered.

"Inhabited by what?" I asked, partially out of an increasingly urgent interest, and partially to make him stop smiling, or smiling in quite that way.

"By what, indeed?" said Wake. "The letter gave some vague hints – to one who knows something of what is possible in these cases – and I gleaned a little more from an old woman here who is a bit of a folklorist and local historian.

"The letter spoke of the moor's evil reputation. Ever since there have been people living on the moor, it seems, the moor has been feared. It is known that young men are especially in danger when they walk alone upon the moor. In fact, women, alone or in groups, may walk the paths in perfect safety, at least from – from whatever it is. This has always, it seems, been known. So says the letter.

"My local folklorist goes further. She says that she has uncovered certain evidences that, in pre-Roman times, the people of these moors would offer young men to – to whatever it is that takes them and drowns them out there. She theorizes that to walk the spiral path alone – there is still a spiral path winding out from the moor's centre – was a rite of passage, a risk each boy must run, for the good of the community, before he could properly be considered a man. Usually, the boy would return. Sometimes, he would not. Then it was said of him that one of the bogs had taken him to its bed."

Wake shrugged. "She admits that much of this is guesswork, but she has partial and suggestive evidence – a legend here, a children's song of chilling antiquity there, and a single reference in a Roman record to a tradition already dying when they came." Wake paused, and there was a new seriousness in his voice when he next spoke. "In other words – whatever's out there, it's old."

"But why so many losses just now?" I asked.

Simon Wake smiled sadly. "There's a very simple answer to that question, though it took me some time to get on the track of it. A few months ago, one M. Atherton published a chatty guidebook entitled *Moorland Tramps*, giving routes of walking tours one can take on various moors, and rhapsodizing about the charms of doing so. There's a chapter on the moors around Blankford, where M. Atherton had a lovely time, walking in glorious solitude. M. Atherton is also (though this fact has been carefully concealed throughout the text of *Moorland Tramps* – I expect because her publishers rather thought it would sell better as a masculine account) a woman. She was therefore safe from – it – on her own tour. Unfortunately, many of the people who read *Moorland Tramps* are men. Therefore, many young men have been coming alone to the moor of late. And some of them drown."

"But what actually happens to these solitary men?" I asked.

"Why, they leave the paths, and walk, through thorns and hedges and over rocky soil, past dire notices of their doom posted everywhere, past all sorts of inconveniences, right into one of the three deadly bogs, where they drown. I expect I almost followed them, last night."

"Ah," said Anderson, looking quietly pleased.

"Yes yes! Get on," I said, impatience and terror warring within me.

"I had two local worthies accompany me out onto the moor, starting at 11:30 at night. At about 11:45, they left me. You see, I'd worked it out that I couldn't possibly get from that point along the path into any of the local bogs in fifteen minutes. And – at midnight—" he gestured, dispelling the smoke of his pipe with his hand.

"Yes. We quite understand," I said hurriedly. "Good thinking."

"Anything happen?" Asked Anderson.

"Not much," said Wake. "But I think it may have been starting when I – ah – left. I heard something. A voice, perhaps. It seemed to call... but it got all mixed up with the sound of the church bells chiming out the hour of midnight. I noticed later that I'd been torn by brambles. I must have stepped off the path for that to have happened, though I don't remember doing so." Wake stood up. "Now – if you gentlemen are ready – we have a certain amount of walking to do. It is early yet. It will be quite pleasant, on the moor."

"No," I said firmly.

* * *

Twenty minutes later, I was on a moorland path, trying to think lonely thoughts, and to forget the rope tied round my waist. This rope extended in front of me until it disappeared over a rise; it also extended behind me until it disappeared in the darkness behind some scrub. This was, it will not surprise you to hear, a dashed nuisance. Still, if it was a lifeline, as Wake rather thought it might be – I supposed I could stand it.

"You must imagine yourself alone. You must try to forget the rope and what it means, even as you keep it from tangling up on things and generally jolly it along. You must forget that it makes us, in some senses, a group, travelling together. This should have – interesting – results." That'd been Wake's little speech as he'd tied the rope round me.

And now, here we were – no, here I was, I reminded myself – lantern in hand, watching the stars above me being slowly swept from the sky by a bank of purple clouds that rose like a colossal monster from the horizon.

I tried to feel the peace that comes of being alone in nature, but all I felt was the fear that comes of being alone in the dark. I forced myself not to try to see Wake's light ahead of me, not to turn and try to make out Anderson's light behind me. Keep to the rhythm of the walk, I told myself, and allow the light of your lamp to describe the whole of your tiny, lonely world. Stepping as slow as a bride, I made my solemn way along the path between the hills.

And then, quite suddenly, there came a great tug from behind and I was sent sprawling. My lantern was jerked from my hand, and I found myself moving rapidly along the rocky path. It hadn't seemed rocky before, mind. But it was very obviously rocky now. I grabbed at a scrubby shrub and tried to hang on. Somewhere from, I fancied, the direction of the westernmost bog came a sound. A... voice? It was faint, but... yes, it was a voice. As I strained my attention to catch the elusive sound, there came another great tug, and this time I was jerked right off the path and into a patch of heather.

"Pendleton?" Came Wake's voice, faintly, but just audibly. "Is that you pulling?"

"Not me," I reported. "Anderson?" I called.

Anderson did not reply. The tugging on the rope intensified. I was dragged through the undergrowth once more. Finally, finding myself in the vicinity of a gigantic boulder, I managed to grab onto it, and then to wrap my body round it, clinging desperately with every available limb. I don't know if you've ever clung with your chin. I did, that night.

And then there came a vicious tug from the other direction. It was so unexpected that I think it must have toppled Anderson. I know it toppled me. Sent me down onto my face in the dirt. Spluttering, I cried out.

"Wake!" I cried. "What on Earth—"

I was jerked forwards again. Then backwards. Then forwards. I wondered, with dreary curiosity, whether I'd be split in two. I couldn't, at the moment, think of a single way that I could stop it from happening.

"Wake!" I cried.

Only silence answered me. Or silence – and, I fancied, voices, calling. But not the voices of my friends. Distant voices, barely audible, calling. One seemed (if I still had my bearings), to call from the western bog, the other – the new voice – from the east. I could make out no words, but the tone was like a coo.

"Waake!" I expected no reply, and got none. Just that terrible pulling. Didn't my friends remember I was between them? Didn't they care what became of me? Didn't they know how it must hurt me? What had happened to them, to make them so heedless? I felt horribly scared, horribly alone, and horribly abraded. Really, it was turning into a miserable evening. Also, I feared to think of what had become of my suit. Ruined, probably. Along with my torso. Tears welled in my eyes. I admit it. I felt very low, there in the dirt, and very helpless, and very sorry for myself.

"Lovely man," cooed a voice – in my ear, yet from an infinite-seeming distance. "Will you not come to me? It is but a little way."

Oh, that voice! It was delicious. I knew that she – whoever she was – spoke to *me*, that she knew me as no-one could ever know me, that she desired me as no woman would ever desire me again, that she was beautiful, and that she was just a little way to the south. I was flooded with knowledge of this. I knew, for the moment, no more.

I must have struggled fearfully in those moments of oblivion. Though Wake was pulling eastwards and Anderson to the west, I fought both of them in my desperation to get to the bog to the south. It was all I wanted. My mind was full of it, and full of her voice, calling me by childish pet names I'd forgotten years ago, calling me her darling, and begging me to hurry. Begging me to hurry became quite a theme, really. If she wasn't so obviously an angel, I might have said she was getting impatient.

The next moment that I was really aware of what was happening was when I came to, so to speak, in a thorny patch of bleak moorland ground – to hear my sweet lady speaking, but not to me, and in a much less beguiling tone than she had so far employed.

"Sister," she called, in a voice that, though beautiful and delicate, yet seemed to shake the ground upon which I lay whimpering, "why are you stealing my prize? He does not come. He ought to be here long since. Which one of you is so vile as to call to two gentlemen at once, when there are three of them and three of us, and when all could be fairly shared?"

"Sister, do you conceal your own crime under this remark? For my gentleman also cometh not, and cometh not. I call, and he yet cometh not. Perhaps it is you who have done this thing," called a voice from the west, a voice like pale apples rotting in darkness, a voice that made the roots under my hands coil like snakes.

"Or perhaps," said my lady, with a voice like home and sweet forgotten things (only angrier-sounding than either), "it is our sister to the east who hath betrayed us both."

"Oh, sisters! Why do you conspire against me in this manner? Why do you seek at once to rob me of my prize and to blame me for being robbed? For my gentleman, also, has not come," came a voice from the east, a voice like dawn light on broken glass, a voice that made the weeds all about me thrash like a sea.

"Perhaps there is some mistake," said my own sweet charmer.

"And perhaps there is none," said the voice like rotten apples.

"Yes," said the voice like light on broken glass. "And perhaps there is

none. We shall test it. Call to your fair gentleman, sister of the western kingdom."

There was a murmur from the west, a sweet sound just discernable as being from the voice of rotten apples. There were, I am sure, words in the murmur, but they were for no ears but Anderson's.

Suddenly I was yanked backwards. Fortunately (in some ways) the rope had got tangled a bit with a fairly substantial tree (for the area, anyway), and I managed, by cleverly slamming into the trunk, to arrest Anderson's progress completely.

"He does not come," said Rotten Apples a moment later. "Call to yours, sister."

"Beloved! My arms ache to enfold you, my lips to kiss you," said my beguiler (among other things – promises that set my blood racing even now, and my cheeks flaming).

When I came to myself again, I found that I'd fallen into a deep ditch. My legs still stirred the mud in which I lay with feeble gestures as of walking. I was aware of a tremendous force holding me back from both ends of the rope. I have trouble believing that I'd moved at all against that, but – well, I must've done.

"He does not come," said my fair one. She sounded – well, beguiling, of course, but also sulky. In fact, she sounded exactly like a fair seductress who has lost hold of her temper and is about to thoroughly disgrace herself. "Call to yours, witch."

From the east came a murmur, only barely recognizable as the voice of Broken Glass. There was a terrible strain on my rope, but I simply lay in my ditch and let Wake pull. I am not a strong man, but I am a fat one. This was, at the moment, definitely to my advantage. The ditch was, as I've said, deep, and he'd have had to lift me out of it to go anywhere.

Eventually, both murmur and strain died into the nighttime quiet. Daring much, I emerged from my ditch. It felt good, to stand on my feet.

"He comes not, sisters. What treachery is this? Why have you turned against our ancient bond?" Cried Broken Glass, in a tone of shrill, concentrated rage.

"Do you address me, sister?" Said my own fair one, sounding really very cross.

"I address both of you." Now Broken Glass sounded dangerously calm. "It is quite clear that you conspire against me. It is a thing I have known for some time. You envy me the glories of that which is mine. Here, the cranberry grows in abundance, and there are many flowers that shine like jewels in the morning glitter. You hate me for these splendors. It is quite natural that you should."

"How dare you, whelp?" cried Rotten Apples in a rage. "How dare you compare your noisome little kingdom with the austere majesty of mine? With my waving grasses and abundance of birds? But you have spoken truly in one matter, trull: there is treachery riding unbridled amongst us. Specifically, it is the two of you who, long envious of me, now seek to rob me!"

"Envy you?" said my sylph. "Pah! I spit on your dim and unpleasant shadows of beauty. In my kingdom dwell all the colours of the rainbow, filling the very air with tiny shining wings. Many funny men with nets have drowned in my kingdom without being called, sacrificing their very lives in tribute to the beauty and the rarity of the bugs that live herein. As for your wet and dripping mires – I would no more have them than I'd have the plague. Nasty, oozing buboes. In fact, are you quite sure that your kingdoms haven't the plague themselves? And how sure are you that your gentlemen die cleanly of the waters, and not of some rampaging contagion? You cry, *oh, come to me—*" (and at this point my legs twitched to obey her, even though her tone at the moment was no more than a mocking shadow of command) "—and they stumble into your stinking kingdoms – to be overwhelmed by the noxious stenches – then – splash! – dead before they hit the water." My sylph sounded highly pleased with her own invective – which, I had to admit, was really very good, of its kind. "Oh, sisters," she said, in a kind of fierce ecstasy, "long have I ached to speak thus to thee."

There was a brief, shocked silence.

"Do you truly assert," said Rotten Apples, "that your bug-infested mire, bubbling as it is with muddy pustules – such that I have often wondered how it dares turn its diseased face to the sun—"

"Not," said Broken Glass, "that it gets much sun – not through that choking canopy of weeds—"

"—That your little mudpit – that your disgusting half-acre of slime – that that miasma-ridden infection upon the fair face of Nature – that *that* is superior to my glorious and sunny kingdom of grass?"

"To be absolutely fair, sister," said someone – and it took me quite a time to realize that it was, in fact, Simon Wake—"your endless and undulating grasses are—" he giggled coyly "—a trifle dull."

"Which of my sisters," said Rotten Apples, "has dared to say this?"

"It was not I," said my sylph. "But I thoroughly agree."

"And nor was it I," said Broken Glass. "But it was well-said."

"Cowards! Traitors! Thieves! Only death shall settle this," said Rotten Apples.

"Only death," agreed my sylph.

"Death only," said Broken Glass.

"Let us meet within the stone circle and summon our gentlemen to us," proposed Rotten Apples. "They shall act as our champions in this matter."

"The stones have long been taken," said my sylph. "But I yet know the way."

There was a great rushing noise, and somewhere in it, I heard Wake shouting.

"Tie yourself down! Quick, before they call us!"

I stared around me, confronted with the problem of tying a knot in a rope that is attached to things at both ends.

"Oh, wait!" I realized. "No, it's easy enough. I just have to walk myself through the turns of the knot, as it were – and then—"

"Lovely man," came the voice of my own sweet sylph. "Come to me."

I knew, for the moment, no more.

* * *

I returned to consciousness halfway up a desolate and rocky hill. Anderson was there, too, blinking stupidly about him. From the hilltop came screeching cries and the sounds of blows. Flashes of light seemed to shake across the cloud-heavy sky, in which monstrous shadows were briefly cast.

"I suppose they started the fight without us," said Anderson. "Managed to get my rope round a pretty sturdy boulder before – before I was called. Suppose it must have delayed us long enough for 'em to fall to fightin' personally."

"Where's Wake?" I asked – and then I looked at the rope that connected Wake and myself. The rope led upwards, towards the crown of the hill.

"Wake must be up there – with – them..." I said, weakly. Anderson and I stared at each other in horror.

The screeching rose to a crescendo then died into utter stillness for a long moment. Then, from the top of the hill, came a terrible – a very terrible – thwack.

A meaty sort of thwack.

A horribly final-sounding thwack. Then, silence. Anderson and I ran up the hill.

The scene that the moon disclosed atop that bare and lonely height was one I shan't soon forget. Two bodies, lean, dog-faced, and horrible, lay upon the ground in the stillness of death. They each had terrible wounds, and they reeked of bog. Blinking against the strangeness of what I saw, I realized that the bog was pouring out of their corpses, as blood might pour from men. As I watched, the corpses were already beginning to sink, enfolded into the embrace of the increasingly sodden earth.

Simon Wake stood over the third figure with a rock in his hand. And its head – its head was the wrong shape. Something had happened to its head. The thing gave a final twitch and lay still. With a sob, Wake dropped the rock. It splashed. Only then did Wake become aware of our presence.

"We'd better get away from here," he said. "I think by morning this hill will be gone. There will be a bog here. With butterflies, you know, and grasses, and little flowers like jewels." He laughed unsteadily. "Down the hill. Now."

We tumbled, still tied, down the hill. Only at the base did we speak.

"Those – things—" I puffed, as I fumbled at the knot about my waist, "were – the voices?"

"She would have looked very different to you, had she gotten you into her bog. I imagine she would have been very beautiful, and even – in her way – kind. And then you would have drowned. Quite happily." Wake sounded vaguely wistful.

"And what were they?" I asked.

Anderson and Wake both looked at me as if at an idiot.

"Fairies, old chap," said Anderson gently. "Obviously – fairies. I imagine Wake would have – ah – a bit of trouble saying so, but – Wake?"

"Yes," said Wake, shortly.

"Oh," I said. We were all silent for a moment. "And you killed one of them with a rock."

"I did," said Wake, staring straight ahead of him.

"Ah."

"She was already badly wounded, I expect," said Anderson.

"Yes."

"She'd killed the other two, but not without being badly hurt herself," said Anderson.

"Very badly hurt. But she might have got better," said Wake.

"Now she won't, though."

"Now, she won't," said Wake.

"Is that – killing a fairy, I mean – going to cause you any – ah – trouble? At home?" I asked.

"It isn't my home," whispered Wake, with incredible bitterness.

"But will it—"

"I haven't an idea. Let's go back to that beautiful, dull little villa of Anderson's, shall we? Tea – and perhaps some of that trout – and—"

We'd begun to walk at this point, with Wake trotting just a little behind us. But now, incredibly loudly considering how far we must have been from any church, came the chiming of bells. And I knew, before I'd counted the twelve chimes, what time it must be.

Simon Wake wouldn't get tea and trout in Anderson's beautifully human, blissfully ugly little front parlour.

Because Simon Wake was gone.

GHOST TRAINSPOTTING

PAUL ST JOHN MACKINTOSH

"Do you like playing with trains?" I asked Quigley, when I called him up from our Edinburgh HQ with details of my latest assignment. "Because if so, I've got a beauty for you."

As Scotland's only insurance adjuster for ghost-related claims, I do get cases that are not only spooky weird, but weird from any standpoint. Crazy, loopy, daft weird. As daft as the brainstorm that led my firm to adopt its gimmick Ghostsafe policy in the first place, offering commercial damages insurance against hauntings and other psychic manifestations. But sanity never dawned, and the policy is still in place. And policyholders still claim against it.

I could hear Quigley practically panting as I outlined the case to him. This time round, the claimant was WaverleyRail, a tin-pot self-styled successor to LNER, LMS, and the other pre-war liveried railway giants, freeloading off the Waverley brand. I sometimes think it's a pity that Sir Walter Scott is out of copyright, with all the idiots taking his name in vain. Anyway, this outfit had sunk hundreds of thousands of pounds into a feasibility study and pilot scheme for a new service in East Lothian, along disused lines from Edinburgh to Prestonpans and North Berwick, terminating at Dunbar. Now the whole project had apparently been, well, derailed by Transport Scotland, and the franchisee was desperately scrabbling to salvage all it could from the scheme. Including a commercial damages claim against reports of a ghost train haunting the disused tracks.

"If it's trains you want, I know just the guy," Quigley slavered. "Dow's a buddy from way back. He's yer man: Ace trainspotter, always bashing the tracks and walking the abandoned rails. He's probably logged every ghost train sighting in the country since the Tay Bridge disaster."

I sighed under my breath. I shouldn't be surprised that Quigley knew a leading railfan. Geeks of a feather flock together. Quigley might be the nearest thing I have to an expert witness in his crank area of pseudoscience, but I never let myself believe for one moment that he was a stable, well-socialized individual. He'd pulled in other twitches on other cases before now, and I was already sure that Dow would be one more of the same.

"Fine, call him up in case we need him," I instructed Quigley, then signed off, still hoping that I could kill this claim before it got to the fieldwork stage.

The claimant was nervous enough when I went to interview them in their Thistle Court offices – though not of ghosts. "Our legal advisors warned us that we were looking at immense potential damage claims if any passengers were harmed by ghostly manifestations," bleated the youngish executive delegated to see me. "That really upset our risk projections."

"Oh?" I asked politely, trying hard to keep the irony out of my voice. I knew that the Ghostsafe policy was usually offered as part of a commercial insurance bundle, and that WaverleyRail would probably never have even noticed it until they started to cast around for ways to salvage every pound they could from their project. I wasn't going to rub that in her face, though. Not yet anyway.

"It wasn't just the reports of hauntings; it was the type of ghosts," she continued. "I mean, wartime trains. With guns. That seemed so much more risky, and scary."

I actually snorted at that. I couldn't help myself. I mean, what were the ghost trains supposed to do? Shoot up WaverleyRail's carriages with phantom shells? The girl looked offended.

"It's all there in the reports," she sniffed, prodding the slipcase on her desk. "You can go ahead and verify it."

I took the file away to pacify her, even though they'd already sent us soft copies of all the materials. The actual eye-witness sightings almost all came from one source: a retired couple occupying a former railway halt on the abandoned coastal line between Aberlady and Gullane, running east from Edinburgh along the Firth of Forth, which had been converted into a retirement cottage. Several local papers and websites, and even one item in *The Scotsman*, quoted their eye-witness account of a wartime steam train, or trains, in olive-drab camouflage, with gun barrels protruding front and back, and helmeted heads peering over the armour-clad sides. Only, its headlamp and riding lights shone with a spectral radiance, and other lights and objects showed through it. It whistled and clattered like any train, but with a muted softness that seemed to carry over a great distance. The old couple had seen and heard the manifestation several times, always at around the same time at night, and always on the same small branched network of lines around their house. WaverleyRail's project, if it had come off, would have brought the whole line back into service. Immediately, I wondered if the couple had made the whole thing up to avoid any disturbance of their peaceful retirement, and I resolved to go down and interview them.

At least the drive out to Aberlady was short enough for me to be there and back in one morning, so I didn't have to disturb Morag and Jennie with another night away from home. Truth be told, I was tired of having to explain my assignments to my wife and daughter, and watch the tolerant patience steal into Morag's gaze, and Jennie's lips assume a sardonic smirk. Sparing my own blushes, I took the pool car east. With WaverleyRail's plan on hold, I couldn't even take the train.

The cottage made me even more suspicious at first. It was less conspicuous than I had expected, set back from the bleak coastline and the view of the Firth of Forth from the A198, but the owners, Mr and Mrs Gifford, had stressed the railway nostalgia theme so hard in their decorations that you could practically smell the steam and soot. They had the old LNER station nameplate and Thirties railway ads for Edinburgh and the Firth of Clyde on their house front, beautifully maintained, and a gleaming white picket fence along the disused platform. Mr Gifford, a genial old buffer in his eighties, let me in, and I saw that the couple were, if anything, even more obsessive in their own quarters. Railway memorabilia filled practically every square inch of the interior: signal lamps, engine nameplates, gleaming brass whistles, timetables, even an entire smoke box door with handles and number still intact. The fireplace was a lovingly converted steam engine firebox. I started to wonder if the Giffords really were that keen to keep the trains away.

"Aye, we both saw it," beamed Mr Gifford, as his wife served me instant coffee and Dundee cake. "And heard it too. Clankin' and whistlin' fit tae bust. Tommies oon board it an' aw'. Chuffed past the hoose more than one time, lights shinin'. Only after dark, mind. It's a ghoast train awreet."

"Just like that Arthur Askey film from the war," Mrs Gifford nodded.

The Giffords clasped each other's hands and gazed at each other lovingly, eyes shining. I started to revise my initial hypothesis from deliberate fraud towards joint hallucination. If they were this keen, obsessive even, then perhaps they wanted the trains back after all – enough to wish them back in their imagination.

"Then you can testify that you did see what you thought was a real phantom train?" I persisted, recording the whole conversation with my little pocket dictaphone.

"Not just any train, laddie," the old man breathed heavily, leaning close. "One of the old Home Guard armoured trains, that patrolled all along this coast, keepin' us safe from the Germans. I could see the cannon barrel pokin' out of her front wagon, and the armour plates on her sides. Manned by the Poles, they were, before the LNER men in the Home Guard took them over.

Back in the days when the railways around here really were something, before British Rail and the Beeching Axe. The crew used to barrack at North Berwick. Armoured Train N, it mustae been, mounting two ex-Navy six-pounders and Vickers guns, bristlin' wi' Brens and rifles. Better than any tank o' the time. Wouldae given the Jerries a nasty shock if they'd ever dared try tae land."

"There's no chance you could have mistaken some other train?" I persisted half-heartedly. "Perhaps some lights late at night? Or a goods train?"

Mr Gifford bristled himself. "I know mah trains, laddie. I looked her up in the gazetteer and all. Could even tell ye her engine number if you want me too." And he brandished his mug of tea towards the shelves of Middleton Press books and trainspotting guides tucked either side of the firebox fireplace.

"And did WaverleyRail talk to you about what you saw? Try to find out more about it?"

"The dunnerheeds couldnae run a train if it were Thomas the Tank Engine," he snorted. "I'm nae surprised they're pullin' oot. Daft Sassenachs, the lot o' them." His wife tutted sympathetically.

I didn't see any point in probing the Giffords any further: whatever the truth of their story, they were going to stand by it, and wouldn't provide any evidence to undermine the claim, no matter how dismissive of WaverleyRail they were. And I didn't want to upset them by drumming in that I was there to disprove their story, and demonstrate that they had seen nothing, so that we could duck the damages claim. I was going to have to give Quigley his day in the field.

Before I left the site, though, I took a quick look down the back of the cottage, where the abandoned track ran. WaverleyRail had progressed no further than their feasibility studies, and hadn't started to actually clean up the track yet. Short grass grew between the rusted rails, along an empty cutting of several hundred metres extent, that led round a gentle curve to a dark, boarded-up tunnel mouth, like a fairy-tale cavern fenced off to keep the fire-breathing dragon at bay. Already it looked melancholy and sinister by the light of day, and I could almost imagine that a ghost train might come whistling round the corner at any moment, even without the Giffords' promptings. I suppose you had to be obsessives like them to be comfortable living next door to such a gloomy, evocative place.

Quigley, though, was over the moon at the whole prospect. "I'll rope Dow in," he burbled, when I alerted him by mobile. "We'll be able to do a complete workup of the whole site: him for the railway stuff, me for the psychic aspects. It'll be one hell of an opportunity. I can't think of anyone ever getting proper readings from something as big as a phantom train before."

"Just try to keep your expenses claims down," I hissed. "We can't run to much extra gear on this jaunt. The site is big enough, and I don't want you trying to festoon it with sensors or anything."

Quigley agreed readily enough: after all, I was giving him the chance to pursue his obsession again, and to get paid for it, and to be taken halfway seriously. So did Beattie, my supervisor, when I rolled back to our head office in the Edinburgh Exchange. "Just get it over with," she nodded. After all, Quigley's findings hadn't yet managed to convince one single court, even when he claimed to have firm evidence of ghostly manifestations. We had to wheel him out as a show of good faith and due diligence, but we never expected him to find anything, and he had never disappointed. No matter what he thought or hoped, he was one of our best resources for stopping claims in their tracks.

Morag was less keen, even though I explained that I wouldn't need to spend the night away from home. She got that martyred look again, and the prospect of me coming back to Blinkbonny Lane in the wee small hours was only partly to do with it. Jennie was more forthright. "Aww, off to play with trains, are ye, Dad?" she sniggered. "Bring us home a spooky Thomas, will ya?"

After that, it was almost a relief to get out to Aberlady the next day. Quigley arrived at the Giffords' place just after me, in his Fiesta hatchback, stocked, as I knew too well, with all the paraphernalia of the modern ghost hunter, from thermal imagers to motion sensors, all bought from Haunted Rooms. He brought Dow with him, and if anyone could make Quigley shine as a comparative beacon of charisma and social adjustment, it was Dow. He looked as though an Oxfam shopper had stuffed a ferret down his trews, which had then devoured him and started impersonating him. I didn't want to touch his anorak for fear it was contagious. He knew his trains, though; I'd give him that.

"Going by the description, it'll be one of these," he mumbled, fingering the screen of a greasy tablet. "Hauled by a F4 2-4-2T tank engine: not one of the most successful designs. Called 'Gobblers' because they ate so much fuel. LNER must have been glad to hand them off to the Army."

He showed me a picture very close to the mental image I had formed from Mr Gifford's description: a string of ironclad camouflaged wagons, bristling with tin helmets and bayonets, firing off a broadside from big shielded guns fore and aft. Then he pulled up another shot, of what looked like a colourful staged diorama of a similar vehicle emerging from a tunnel, with dummy Tommies leaning over its end. "That's from the Bovington Tank Museum," he proclaimed proudly. Quigley, craning over my shoulder, sighed in fond accord.

"So historic," he breathed. "Just like the whole line. I can see why your claimant was so eager to set up a service along here."

"Historic?" I asked.

Quigley turned a pitying gaze on me. "Aye, historic. Go a little south, and you've got Athelstaneford: That's where Scotland's banner comes from. According to the legend, St Andrew came in a vision to Oengus, King of the Picts, promising him victory against Athelstan's invading army of Northumbrians. The next morning, Oengus's Pictish army saw the clouds above the battlefield form a white St Andrew's cross, and under that sign, they triumphed and slew Athelstan. The St Andrew's cross has been Scotland's flag ever since."

"Well, fine," I shrugged. History isn't my thing: I have enough to worry about in the present. It turned out to be a common bond for Quigley and Dow with the Giffords, though: pretty soon they were prattling happily in the forecourt of the old halt, about the golden age of the North British Railway and the building of the Forth Bridge. I let them get on with it, watching the sun dip towards the trees at the western end of the cutting. I knew from experience that Quigley wouldn't need too long to set up, once he had the location mapped out.

Mr Gifford decided to join us for our evening vigil: Mrs Gifford made sandwiches, and brought out thermoses of hot soup and coffee. At least if I was going to be roped in on a wild goose chase again, I might as well do it in comfort, I reflected. Mr Gifford and Dow went pottering up and down the line, as Quigley rigged up his monitors and instruments, covering the track from both sides. "The lasers will trigger the cameras and sensors, when it passes through the beams," he explained. "And if that doesn't work, I've got a remote control for the whole network. Once it comes steaming past, I'll have it."

"Belching ectoplasm, no doubt," I sniffed. Quigley winced, and gave me a straight look.

"You've no soul, man," he sighed.

"At least I won't be around after I'm gone for you to chase me, then," I shot back. The chilly vigil was getting to me. And, I had to admit, the

Giffords' description and the images of the armoured train had got to me too. That darkening culvert, with its black tunnel mouth at one end, half hidden round the curve of the track, and its forlorn abandoned air, seemed just the right setting for a ghost train to come puffing and whistling out of a grave long enough to hide an entire train in. I've made a career out of my scepticism, but if you're forced to hang around with flakes like Quigley for long enough, you can't help reacting the same way and triggering on the same things they do.

Finally, the last light faded from the sky, and we sat down together on the cold station benches at the back of the Giffords' house to await whatever might come. "It's like the ghost ships of yore," muttered Quigley, huddled over his thermos. "And the phantom riders and ghostly coaches. Or even the curse of James Dean's car, Little Bastard. All the most romantic, evocative vehicles get projected into the afterlife. Like Abraham Lincoln's funeral train. Or Stockholm's silver subway carriage. Or the ghostly bombers over Loch Lomond. Amazing the power of the human spirit."

"If you say so," I shrugged. I was shivering, from the cold of course. Mr Gifford had kept most of the old platform lights off, so we sat in almost complete darkness, with just a little yellow glimmer from the curtained cottage windows. Far off, there was a gentle sighing or rushing in the air: either the wind in the trees, or the distant sound of waves in the Firth of Forth, I couldn't be sure. I jumped when an owl screeched from the trees over the cutting. Just startled, of course.

"Listen," Mr Gifford hissed, from the bench nearest the tunnel mouth. I listened. Quigley swivelled his head like a pointer sighting a downed grouse, and immediately brandished his remote. Dim green laser beams shot out across the track. Even so, I still couldn't hear anything... until I made out a faint modulation to the rushing sound in the air, a rising and falling susurrus. Our little party rose as one. I'd already worked out that it was the almost panting puff of a steam engine, carried over a great distance. Then came the far off note of a whistle, high and shrill, yearning and poignant, like a banshee's keening over the dead on a remote battlefield.

"That's it," Quigley hissed. "Here she comes." We moved up the culvert a little way, but stayed on the platform. We were all now scared as well as expectant, I felt, no matter how eager the others seemed to be, and I noticed that we all edged away from the track. I could already picture the ghost train chugging and clanking past us with the dim lights showing through its sides. Quigley's cameras and sensors should capture it, thank God, so we could stay well back from it, and even close our eyes if it got too much.

The whistle sounded again, stronger now, echoing as if from off the tunnel walls. The pulse of the steam engine was now unmistakable, resounding and reverberating. At the far end of the cutting, I could see lights flickering and flashing between the trees. I fought down the urge to grab hold of Quigley's shoulder. Dow and Mr Gifford were actually clinging to each other as the noises grew, bearing down on us. Then suddenly, through the branches, I saw the headlamps of a train emerge from the tunnel mouth – and steam away to the left, between the trees.

"Damn, I forgot aboot the siding," Mr Gifford cried, and he clambered down from the edge of the platform and staggered down the track, towards the lights and noise. We chased after him. We could still see the lamps shining among the trees, and the outlines of slab sides and even tin helmets, and a spectral cloud of steam belched from the puffing funnel, but they were all passing and fading. By the time we reached the tunnel mouth, they were all gone, with a last faint farewell whistle ringing in our ears.

Mr Gifford was gazing down in exasperation at the track. I saw he was standing next to a set of points that I hadn't noticed before. The diverging siding was even more overgrown and rusted than the main line, with whole sections missing, and led to a gap in the trees that had also been planked over like the tunnel mouth, with a solid hoarding. Squinting past it in the darkness, I saw a featureless empty space that must have been a marshalling yard back in the days when the line was still operational. It had never occurred to me to come up this close to the tunnel mouth to check.

"I didn't think to cover this section of the track," Quigley sighed. "I just assumed the train would come down the main line. But then, ghosts don't take any notice of barriers, do they?"

Disconsolate, he started to take his sensors down. Obviously, he wasn't expecting any further manifestation tonight.

"Well, gents, at least ye saw something," Mr Gifford declared. Dow hung his head. Quigley looked about to make some retort, but held it back and said nothing. I knew from past experience that he needed time to analyse his recordings, and just get over his disappointment.

I thanked them all, got in my car, and drove back home to Edinburgh. Morag was asleep when I got in, as I expected, and I didn't wake her to tell her what had happened. I still wasn't sure myself. When she and Jennie asked me over breakfast, I just told them that we all had seen some lights and heard some noises, and that was all.

Quigley couldn't add much more to that when he called me the next day. "Just a sound recording of some far-off train noises, and a few dim lights between trees," he sighed. "Could have been anywhere. Nothing conclusive." He was all fired up to go out there and try again. "All I need to do is reposition the sensors," he urged. "Then I can document the whole thing."

Beattie was reluctant to pay his fee for another day, though, and while we were still arguing about it, the claimant's whole case collapsed. WaverleyRail had sought insolvency protection, and the receivers couldn't be bothered with a silly damages claim against ghosts. Quigley was disappointed, of course, but still eager to do something at Aberlady.

"I don't mind doing it by myself, unpaid," he panted. "It'll be the psychic find of the century. A whole ghost train, and a wartime one too. I'll be famous."

I felt a little bad for the Giffords, having landed him on them, and I drove out again to thank them for their help, and to make sure that Quigley wasn't making too much of a nuisance of himself. I found him fussing around the tunnel mouth, setting up sensors, well away from the cottage. Mr and Mrs Gifford were out on the abandoned platform, gazing up the line towards him with a kind of fond bemusement.

"He's nae bother," Mr Gifford assured me. "Hasn't seen a thing, but still hangs up his little wires. Had a couple of his friends here too, though they soon went away, including that wee trainspotter mannie. It's the most excitement we've had in years."

"Well, if he is any trouble, just let me know, and I'll have a word with him," I replied.

"Och, it's good to have people takkin' an interest," Mr Gifford beamed. "Paying some heed to the old defenders. And I'm sure he'll see them again some day. If the British Isles are ever threatened again, the armoured trains will ride out to shield and defend us all." And the Gifford clasped hands and smiled into each other's faces once again.

It's a comforting belief, I reflected, as I pulled away from the halt. I suppose we all need our consoling, romantic illusions. I mean, what are the alternatives? Dow, obsessively filling his days with his data books, track bashing, in an activity that's a byword for futility? I'm not the one to say.

I just adjust.

ANGEL SCALES

BRANDON BARROWS

The smell was the worst of it.

I've seen enough blood that it only concerns me if I step in it, and I've seen Enochian scripture scribbled in far worse things in far better handwriting on the walls of apartments a lot less luxurious than this place. None of that bothered me. But the smell did – sulfurous, with a greasy tang that crept into your nostrils and down your throat to coat your tongue. The kind of rank stench that wants to stay with you your entire life, something more than sense memory, something less than natural. It was all I could do to keep from choking, to keep from trying to spit it out. I knew neither would do any good and it wouldn't help my professional reputation any, either.

My companion had no such concerns.

Yumiya was hunched in the corner by the entryway, one shoulder pressed against the wall, retching all over his shoes. I guess he didn't get a chance to change into slippers before it all hit him.

"You okay?" I asked.

The other man, red-eyed and shaky, glanced over his shoulder at me then turned away again. "How—"

The building manager pulled a handkerchief from the back pocket of his slacks and took a moment to compose himself. He was wincing and almost, but not quite, panting when he stepped from the corner and looked my way again. "Kuromori-san, how can you be so... calm? And what in God's name is that—" He waved a hand around helplessly.

I shrugged. "It's my job. Trust me, I've seen worse." That was only half a lie. I've *seen* worse, but never *smelled* anything this bad.

"As for the smell..." I strode across the hardwood floors, taking a slightly circuitous path to avoid still-sticky splotches and little puddles, both of blood and something else I'd rather not think about. There was too much blood to have come from one body; that did bear some thought.

In the center of the room, surrounded by a ring of waist-high sconces holding suspiciously-yellow candles, some still flickering, lay a once-white sheet, now stained with blood and burn marks. Gilt-edged pages torn from an English-language book were arranged around the edge of the cloth in a rough, secondary, inner circle. I pulled the cloth back, revealing the source of the smell, at least some of the blood, and the centerpiece of the ritual: a partially-burned goat corpse, its throat raggedly cut, its horns adorned with the same likely-human-tallow candles held in the sconces.

Yumiya's body tried to reject what it saw, violently, but his stomach was already empty. He turned a shade of green not generally seen in a human complexion, and dry-heaved. He whirled on his heel and leaned forward, clutching his belly and making rasping sounds until the feeling passed.

"Sorry. I should have warned you, Yumiya-san."

Back still turned, he waved that away. "What in heaven..." His voice cracked a little.

"You're not far off." I tossed the shroud back down to cover the mess on the floor. "It's a burnt offering, like in the Christian Bible. Whoever did this wasn't bright enough to skin the damned thing first, though. It's the hair you're smelling. That and the rest of this." I gestured vaguely, though he couldn't see it. Nothing as ordinary as burnt hair, regardless of the animal, could make a smell this vile. I wasn't entirely sure yet what happened, but the pieces left in plain sight were hard to misinterpret.

I glanced around the apartment again, trying to estimate its size. Just from the main room and the doorways branching off from it, it must have been at least a hundred square meters. The flooring, the areas not covered in filth, shone in the early morning light that seeped through half-drawn curtains, giving off a kind of inner glow you only see in the very best wood. There were no furnishings in this room, unless you counted the goat and the candles, but I could see possibilities here. Ichibancho House was one of the most expensive, most exclusive luxury apartment buildings in Chiyoda Ward, the kind of place where people buy, rather than rent. This apartment probably cost two-hundred-million yen, more than I'd see in a lifetime. It was too bad; I doubted it would sell for that much the next time it changed hands. I clicked my tongue in annoyance. Rich, bored and stupid was a terrible combination.

While I was ruminating, Yumiya screwed up his courage and now stood beside me, looking down at the gore-stained cloth. Despite the puking, I was impressed. The man's clothing was ruined, stained with vomit and fear-sweat, but he pulled himself together. There was a job to do and he was ready to face it.

"This is... I don't know." He shook his head.

"Who lives here?" I pulled out a packet of cigarettes. "You mind?"

Yumiya shook his head. "We don't allow – oh, what difference does it make now? Go ahead."

"Thanks." I slipped out a cigarette, held it between my lips. I quit years ago. I didn't know why, on the way to Ichibancho, I felt compelled to buy them. Something extrasensory, perhaps. "It helps with the smell." I offered him the pack, but he declined.

I lit up and took a drag. The sensation was pleasantly familiar, reminding me of how insidious the things were. "Who lives here? And why call me, instead of the cops?"

"You're not...?" The other man gave me a look full of questions. That was interesting. He called me without knowing who I was. I wondered who told him to do that. Time for that later.

"Well, it doesn't matter." Yumiya put his back to the remains of someone's hard work, positioning himself so he could still face me at an angle. "I called you, Kuromori-san, because a certain resident gave me your name and number."

I raised an eyebrow.

Yumiya took the cue. "I'm sorry. We take residents' privacy very seriously here. I can ask, later, if you'd like, if it'd be all right to—"

"Forget it," I told him. "What did this resident say?"

"All he said was, 'Azuma Kuromori' and your phone number..." He cleared his throat. "I thought, I mean... I just assumed," Kamiya continued, "that you were some acquaintance in the police who might keep this quiet. I value this person's insight greatly. I called seeking advice and... I'm getting off track again, aren't I?"

I told him it was fine.

"To your question, then: Fumiaki Chigara-san lives here. And before you ask, I can't tell you anything about him other than he purchased the apartment two years ago and I rarely see him. I can tell you that the doorman and the watchman on duty last night had numerous complaints, both about noise and later about the smell, and that Chigara-san couldn't be reached on the phone nor was there any answer at the door. They called me, I arrived, had the same experience, called my acquaintance for guidance and then you right after."

"I see. And did anyone enter the apartment before us?"

He shook his head. "Not as far as I'm aware."

"Well, then." I took a long drag from the cigarette, burning it down close to the filter, then flicked it onto the shroud. "Sorry. No ashtrays."

Yumiya had no comment. Instead, he asked, "Do you know what happened here, Kuromori-san?"

"I've got a pretty good idea. Your Chigara-san, or someone with access to

Occult Detective Magazine #8

his apartment, anyway, attempted a summoning. I'd guess he succeeded, too, to some degree or another."

Yumiya's face went through a series of slow contortions. Finally: " 'Summoning?' You mean, like... demons?" The façade of polished worldliness fell away and a nasal Sendai accent flavored the question.

On the far side of the room, I threw open the curtains covering the door to a wide balcony. The door was the sliding kind and wasn't quite closed. There was a gap of maybe six centimeters between the door and the frame, not enough to help with the smell, but enough to rustle the curtain. I noticed it when I set foot in the room, but other details seemed more urgent then.

"Not quite." I gestured for him to join me as I stepped outside. Out on the balcony, I pointed towards the pile of shed, blood-stained clothing heaped at the edge of the balustrade. "Summoning as in *angels*."

* * *

It would be untrue to say there is no such thing as demonic possession, but it's rare. Demons get very little out of it. A soul forced to obey doesn't belong to them any more than a stolen car belongs to a joyriding kid. They may be able to use it for a time, but ultimately, they have to give it up and it's usually more trouble than it's worth. Possessions *do* happen – more often than you'd think, actually – but typically the culprits are ghosts or certain kinds of *yokai* who do it for a specific purpose or sometimes simply because they enjoy mischief. Most demons would rather make a pact with a human being, wait out their contractor's lifespan and collect their fee in due time.

Angels are much less patient and far more short-sighted. They are eager to appear on Earth, eager to do 'God's' work, eager to bask in the glory they think reflects on them in the human world, and they don't much care how it gets done. They all have a Messiah complex. They all feel that they are the one who will bring God's divine light back to this plane. A single human life, used in the pursuit of that goal, is meaningless to them. None of them seem to realize they're just one more visitor from the Other Shore.

That makes them all the more dangerous.

* * *

"I really don't think I understand," Yumiya said after a pause of several moments.

"You don't need to. In fact, be glad you don't." I ducked back into the apartment, Yumiya on my heels. I turned to the other man. He stood in the doorway of the balcony, transfixed by the scene. The courage he found was draining rapidly from him. There was no real danger here anymore, but this wasn't the nicest first contact with the Other Shore. I hoped he would never have another.

"I'll see what I can do." I pulled cigarettes from my coat pocket, but thought better of it. "I'll tell you now, I don't know what I *can* do, but Chigara-san needs to be found, at the very least."

There was palpable relief on Yumiya's face. "Y-yes. Thank you." His head bobbed up and down in a nervous kind of bow.

I turned to go, but stopped at the sound of my name.

"What should we do about... this?" Yumiya gestured to the room.

I shrugged. "Find a cleaning service you can trust to keep their mouth shut, I guess."

Yumiya clearly didn't like that answer, but said only, "Let me walk you down."

* * *

We exited the elevator into a spacious, marble-floored lobby. Across the foyer, a maroon-uniformed watchman stood at the security desk and motioned towards us. He had the tact not to shout across the space, but it was obvious that he wanted our – or at least Yumiya's – attention badly.

"Yes, Suzuki-kun?" Yumiya's clothing was ruined, but the minutes in the elevator did a great deal to recapture his composure. Aside from the clothing, he was again the consummate facilitator I first met half an hour earlier.

The doorman flipped around the monitor of the computer at the desk, positioning it to face the two of us. "I think you should see this, sir."

The screen displayed the front page of a *matome* site, a website aggregating links to other sites' articles and videos. The top headline read "NAKED WINGED JACKASS AT TOKYO MIDTOWN HIBIYA! SAD JOKE, NEW TV SHOW, OR THE SECOND COMING? LOLOLOL." The headline, linking to the article itself, was purple, indicating it had already been viewed. Suzuki glanced at the two of us, making sure we were watching the screen, then clicked the link. It redirected to another website and a video began to auto-play.

Shaky cellphone footage showed a crowd, gathered in what looked to be the sixth floor garden level of the huge Tokyo Midtown Hibiya structure, the area they used in all that advertising when the place first opened. The camera swung upwards, reaching a level just over the crowd's heads to show a nude man, large white wings sprouting from his back, standing on the railing of the balcony overlooking the garden. He was gesticulating wildly with his arms, the wings occasionally flapping slightly as if to emphasize some point he was making or maybe to help keep his balance on the narrow, metal railing. The camera zoomed in on the man, giving us a clear look at his face. Distance and the sound of the crowd commenting and questioning, both each other and what their eyes told them, reduced the man's monologue to background noise, but his expression was clear. The only way to describe it was *intense*, as if he truly believed that the words from his mouth were the most important that anyone would ever hear. There was a kind of nobility in those features, as well, but the fervor he whipped himself into swallowed it up.

"That's Chigara-san," Yumiya said quietly.

"Yeah." Suzuki nodded. "Just watch, though."

The winged man's speech was wrapping up. His chest swelled with a deep breath and then, for the first time, we could hear him clearly. In a voice like the ringing of some monstrous bell, in a tone both melodic, ominous, and inhuman, he cried, "I say unto you, good people, you have been told of the Coming, so go now! Spread the Good Word and live as He commands and soon, we shall have the paradise on Earth that He so dearly wishes for us!" Then he spread those immaculately-white wings and leapt into the air, disappearing from the camera's view in seconds.

Yumiya turned to me expectantly.

"Huh. At least I know what he looks like now." I gave the other men a nod. "Well, see ya. I'll be back to discuss my fee when this is over."

All I had to do was earn it.

* * *

Tokyo Midtown Hibiya is a massive shopping complex, completed in early 2018, in the heart of Chiyoda Ward. Overlooking Hibiya Park, it's something like two-hundred-thousand square meters of upscale shopping and entertainment that reportedly serves an estimated thirty-some-odd-thousand people a day. What better place in Chiyoda to 'spread the Good Word?'

All Chigara's visit did, though, was spread chaos.

When I arrived, the place was swarming with police and news crews and curious onlookers who either saw the hundreds of videos that went viral or heard from one of the official news sources that picked up the story almost immediately. Chigara was long gone and with so many people around, so were any chances I had of picking up his trail.

I stood across the street from the south-side entrance of the mall, the closest the police would let anyone approach, and scrolled through video round-up sites on my phone. There were more videos of Chigara popping up by the minute. Not all of them were from T.M.H. In theory you could probably track Chigara's progress through the city, but there was no good way to sort them out in the time I had and no sense in trying.

I closed the browser on my phone and opened my contacts. I swiped across a certain name and waited for the call to connect, keeping an eye on the sky just in case Chigara decided to do some fly-by proselytizing.

After four rings, the call connected.

"Azuma-san? I got your bill, but I told you, I'll need some time to—"

"Sasai, listen, I need your help."

"Still no honorific, huh?" Annoyance tinged the other man's voice.

"Pay what you owe and then I'll consider treating you like a human being. For now, shut up and listen. I need a way to reliably track an angel."

"Huh?"

Yuta Sasai wasn't rich, like Chigara apparently was, but he was bored and slightly stupid. I once referred to him as Japan's worst amateur sorcerer because his dabbling in magic got him into more trouble than he could handle more times than I knew. I got him out of the most recent fixes he found himself in, but he still hadn't paid for the suit that was ruined in the process – or the medical bills I incurred while ruining that suit.

"You heard me. An angel, like from Christian Heaven. Yahweh, the Bible, all that. It's really just another visitor, but I've got a client in a situation and I don't have time to track him down the old fashioned way."

There was silence on the line for a moment. Then: "If I help you, will we be square?"

I rolled my eyes. I'd probably never get the rest of the money out of him, anyway. "Yes, sure, fine. Can you do it?"

"Tell me where to meet you." There was a note of triumph in Sasai's voice.

* * *

Saint Francis Xavier Church was a twenty-minute walk from T.M.H., through the edge of the park and into Kanda. The Imperial Palace was just visible in the distance as we exited the park-grounds.

What a strange world this city is, I thought. Just a twenty-minute walk from the most modern shopping center in the country to a Church venerating a foreign saint dead five-hundred years, with a little detour past the home of the living descendant of a god. No big deal.

The church was impressive: a huge, tan brick building that looked like a piece of European history plucked from its foundations and dropped into foreign territory to spread the Gospel. Which is exactly what it was, which was why we were here. The church was just a building, but it had a lot in common with the being I needed to find.

Sasai was excited when I told him over the phone where I was. "That's perfect," he said, promising to meet me as soon as possible. Thirty minutes later, he showed up and led me here. "Trust me, Azuma-san, this is the best place you could hope for."

"Let's just hurry it up, okay?" I could only imagine what the devoted of Saint Francis Xavier thought of the goings on across the park, but I didn't need any more trouble and time was running short. I still had no idea how I was going to deal with Chigara.

"Sure." Sasai's gaze was cool. I guess he expected me to be more impressed. I'd be impressed if his plan worked.

In the shadow of the building itself, beneath a set of huge stained-glass windows, Sasai said, "This is probably fine." He fished a small, glass vial from inside his lightweight sport-coat. Holding it up to the light, he examined the fine, gray powder it held. "You know, this cost me a lot more than your ratty old suit did."

I took the bait. "What is it?"

Sasai grinned. "Ashes from the bones of an honest to God saint – an Italian sorceress named Galgani, who spoke with angels."

I shook my head. "I don't want to know where you got that. I just want to know if it'll do the trick."

Sasai shrugged. "We'll see. Hold out your hands."

I did and Sasai dumped half the contents of the vial into them. The other half he dumped into his own palm, spit into it, then rubbed the powder into a paste.

When, without warning, his hand flashed out to scrawl a 'T' on my forehead, I slapped it away. "What the hell?"

"Language. Remember where we are."

I sighed and resigned myself to whatever came next.

The other man clasped his hands together, bowed his head and began to mumble in the disjointed, tumbling syllables of the Enochian tongue, inserting the familiar Japanese sounds of Chigara's name into several places, then repeating the entire sequence.

I was beginning to lose patience, thinking this was a waste of time I didn't have, when pain lanced through my skull and Sasai's eyes flew open. To me, he said, "Now! Toss those ashes into the air!"

The words were just a smear of noise through the pain, but I understood enough to comply. I threw the handful of grit over my head and instead of simply falling back to the ground, the individual grains began to glow and dance, like dust motes in a beam of sunlight. They swirled through the air a moment, then came together, converging into a single luminous mass that hovered at eye-level for an instant before crumbling to the ground, again just the remains of burned bones.

The pain in my head slowly receded. My vision slowly cleared. Sasai was looking at me, concern on his face.

"Azuma-san! You okay? Did it work?"

Trying to catch my breath, trying to sort through the images now lodged in my brain, I nodded. "I know where he is."

* * *

The crash of breaking glass and the scream that followed answered Sasai before he could even ask the next logical question.

"C'mon!" I didn't wait to see if Sasai would follow before I dashed around the corner of the building. Sasai's spell worked, but it was unnecessary. Coming here hadn't been a waste of time, though. The vision I saw was of Chigara, more twisted and less human than ever, crashing through one of the stained-glass windows of the very church whose shadow I stood in.

I could have kicked myself for not thinking of it without the aid of Sasai's sketchy magic. Where else in in this part of Tokyo would such a visitor find most comfortable?

Through the churchyard and up four thick, stone steps past a statue of the Virgin and Child, I crashed through a bright red door and into the church proper. I was greeted by the sight of white marble pillars, long pews, and a narrow center aisle to a nave with a low pulpit supporting a pair of bronze angels – and one of flesh and blood.

The thing that was once Fumiaki Chigara crouched on top of the wide lectern between the two statues, beneath a broken window twice the height of a man. Brilliantly-white wings spread out behind him like a scene from one of the stained-glass windows, the effect somewhat spoiled by the splatters of blood that clung to the feathers and the rents in his flesh – from crashing through the glass or from the transformation? I wondered.

Those things were nothing compared to his face. It was recognizably the man from the videos, but even at a distance, the changes were obvious; the lines and planes were sharper, more rigidly defined, more beautiful, but less human. In humanity's place was something Other, something some might call Divine. It wasn't all that different from anything else from the Other Shore, though – not as far as I was concerned.

The two of us weren't alone in the church. A woman in nun's habit bowed *dogeza* fashion before Chigara. A middle-aged man in clerical garb stood a few steps behind her, his back to me. The creature seemed to take no notice of either of them. Its head twitched and jerked as if listening to some far-off sound.

The priest whirled towards me. The expression on his fear was pure terror. "Sir—"

"Father," I moved deeper into the church, keeping my steps even and my voice calm, "take the sister and leave as quickly and as quietly as you can."

He swallowed hard, Adam's apple bobbing, then bounced his head up and down. "Y-yes, that's for the best, isn't it?" He didn't question who I was or why I was giving orders. He was just glad someone was taking charge. He stooped, looped his arms around the woman's shoulders and half-lifted, half-dragged her away, murmuring softly to her the entire time. In her stupor, she was easily led.

'Chigara' let out a high-pitched trill and turned towards the retreating clergy.

"Hey!" I shouted.

His head turned. My heart caught in my chest as his eyes found me. I was wrong: Chigara's humanity was gone, replaced by something Other, but not quite like anything I'd seen before. His eyes were golden orbs, bright as the noon-day sun, and there was something swirling inside of them, aching to crawl out – or drag me inside.

"Azuma-san!"

I shot a glance behind me. Sasai was inside the church and whether he knew it or not, his shout saved me, calling me back to myself. The feeling I was left with was familiar and terrifying, something from childhood nightmares. Even so, a part of me wanted to look into those eyes again.

"Stay back." I waved Sasai off.

"I can—"

"Stay the hell back!" I roared.

'Chigara' didn't like that. It hopped from the lectern and took a few steps closer before opening its mouth, releasing a musical sort of sound that was both beautiful and hideous. Like the feeling in my head, I knew it, too.

"Angel scales," I mumbled. The literal music of the angels. Like the Greek sirens, their voices were able to hypnotize – or kill. "Sasai, get out of here and keep everyone else away, too."

I could almost feel the conflict coursing through Sasai. Maybe he was a crappy sorcerer, but he wasn't a bad guy. He clearly didn't want to abandon me. I appreciated that, but one life – Chigara's – was already lost. I couldn't take a chance on losing any others.

"That sound, it's—"

"Go."

There was a moment of hesitation, then: "Do your best, Azuma-san." I felt Sasai's presence retreating.

So far, the Chigara-thing was remarkably cooperative, but that couldn't last. It opened its mouth and I prepared to resist the effects of the angel scales and their hypnotic power, chanting *norito* softly to myself just to create a kind of white noise.

But when 'Chigara's' mouth opened, it surprised me: in a voice as beautiful and delicate as the finest bells, it asked, "Why? You are children of God... yet I bring you the Word and this is how you act?"

It seemed honestly confused. Angels are single-minded. They all truly believe they are doing God's work. The chaos it spread, the life it stole – all meaningless to it. It was like a child trying to please a parent that didn't know it even existed. If I didn't know better, I'd have felt sorry for it. But I did and I knew if I didn't stop it here and now, the measures it would take to please its god would only get more extreme.

"Because you don't belong here." I didn't want to talk to it. I don't know why I did.

It didn't do any good.

The Chigara-thing cocked its head at me, then unleashed a sound that had nothing beautiful or melodic in it. The noise ripped through my brain like a blade through a ripe melon. I could barely keep my eyes open, but if I wanted to live, there was no choice. The angel thing moved rapidly towards me.

I dug into my pockets, searching for something, anything, to give me a measure of protection, a way to fight back. I knew enough about angels to recognize them, but had never actually encountered one before, and had no idea how to fight it. Knowing its name would be helpful, but short of asking, there was no way of finding that, either. *Kagutsuchi-juu*, in its holster at the base of my spine, was always a last resort, but it wasn't that long since I last used it. I simply didn't know if I had enough spiritual energy to make it work properly.

And I had no more time to think about it before the screeching creature was on me.

With a sort of hopping skip, it took to the air, a single flap of its wings pushing it forward and onto me faster than I could even see. Fingers strong as the hand of God Himself flashed out and latched onto my throat, lifting me off my feet then pulling me down until we were face to face. I was taller than 'Chigara', but the thing's strength and presence seemed to dwarf me, making me shrink back into myself.

'Chigara' opened its mouth and let out another trill of melodic horror. The otherworldly sound carried with it a kind of raw emotion I'm not sure I've ever seen from a visitor before – desperation. Waves of grief and sadness washed over me, filling up my head and my soul with this thing's desire for a better world, and for an instant, I was ready to succumb, to let myself believe that it knew the way to that wonderful world. It wasn't merely an invader from the Other Shore, it truly was a messenger from a divine being that wanted to cradle our souls in its embrace, to bring us to everlasting paradise if only we'd let it, if only we'd listen, if only—

"Azuma-san!"

For a second time, Sasai's voice brought me back to reality, but it was a close thing and for a moment, I couldn't understand why he was even there.

It was enough. The Chigara-thing's attention was divided and its grip loosened on my throat.

Sasai stood on the edge of one of the pews, his fingers held before him in an unnatural, painful-looking configuration, chanting rapidly in the back and forth cadence of the oldest forms of Shinto liturgy. I knew both sign and words well. The angel-creature recognized them, too.

"Paganism!" it shrieked, its beautiful voice rising an octave. "In the house of the Lord!" I was forgotten in its outrage and fell from its grasp as it moved toward Sasai.

Throat raw, I gasped for breath, but there was no time to catch it. I fumbled in my jacket pocket for a *fujikome* – containment – *ofuda*, leapt to my feet and slapped it squarely between the thing's wings. It roared and whirled, knocking me aside with a sweep of those great, white wings, before turning back to the still-chanting Sasai. It seemed more insulted than pained by my attack.

"Run!" I screamed, throat raw, but Sasai didn't need to be told. He turned and skipped to the side, dashing along the narrow length of the pew to avoid the thing's reach. It opened its mouth and screeched, unleashing a musical outrage that shook the building, threatening to bring the stone pillars of the church down on all of our heads, before taking to wing again, closing the gap between itself and Sasai. Sasai tried to duck, but slipped and stumbled, crashing down onto the pews. It probably saved him – the Chigara-thing sailed right through the space where he should have been.

I watched, helpless, as Sasai scrambled over the pews, barely avoiding the reaching fingers of the 'angel' that stole Fumiaki Chigara's life and body. I was at a total loss as to how to deal with this thing. The *ofuda* I tried did nothing but annoy it, and that seemed more from personal offense than anything spiritual or supernatural. I had one option left at my disposal, but only if I could summon up the inner reserves to use it.

My heart raced in my chest as my hand found the smooth grip of the *Kagutsuchi-juu*. To the unknowing eye, it looked like an old Imperial Army-issue sidearm, but the Type-26 revolver had been modified nearly a century ago in a way no gunsmith would ever dream of – the intricate series of characters on its barrel were too small and too finely-engraved for any attempt by ordinary hands.

"*Hi de yami wo harae*," I began, mumbling the beginning of the chant as quickly as I could, feeling the rising tide of energy in the very center of my being, hoping it was enough. I was so engrossed in the process that I never saw 'Chigara' until its feet slammed into my chest. The breath *whooshed* from my lungs and *Kagutsuchi-juu* flew from my hand as the blow knocked me back far enough to clear two entire pews. I landed in a heap, slamming my head into the polished stone floor of the church.

Blackness threatened to close out the sight of whatever came next, but I wasn't so lucky. 'Chigara' was lifting me back to my feet. It snarled in my face, its breath strangely cold and oddly sweet-smelling, its eyes two bright, burning pools of crazed devotion. The last vestiges of Chigara were gone, replaced entirely by the usurper.

"You children, you children…" it huffed. "I bring the light and you wallow in the dark—" It choked on its own anger and the expression on its face twisted further, giving Chigara's features a grotesque slant that would have been nauseating if I wasn't already so out of it. The back of my head was cold and felt pulpy, my vision swam, and senselessness crept closer, but I could still see it before me, crushing my throat, tighter and tighter, wheezing out a mock-Biblical tirade against the unworthy and ungrateful. I could think of better ways to die, but I tried to comfort myself that at least it would be over soon.

The darkness washed over me and I began to go under – and then there was a sound like the crack of thunder, but disturbingly close. Wetness splattered my face and I wondered for a moment, *was it supposed to rain today*?

And then I was lying on flat, hard coolness and someone was gently slapping my face and calling my name. "Azuma-san! Azuma-san, can you hear me?"

I opened my eyes. Sasai stood over me, concern on his face and the *Kagutsuchi-juu* in his hand. A tendril of smoke curled upwards from the barrel.

"What?" I croaked.

"Oh, thank God," Sasai said, true relief in his voice. He helped me into a sitting position against one of the marble pillars.

"What did you...?"

I let the question die unfinished. The thing that was once Fumiaki Chigara lay a few feet away, face down in a rapidly-spreading pool of thick, black blood. The back of its head was missing and the luster of its wings was fading before my eyes. Something cold and wet lurched in my stomach.

Sasai held *Kagutsuchi-juu* up. Softly, he said, "It's still a gun."

"Right." That was all I could manage.

Sasai helped me to my feet, then offered *Kagutsuchi-juu* to me. "We make a good team, huh, Azuma-san?"

I accepted and re-holstered the weapon.

There were sirens in the distance. The parallels to the last time I saw Sasai were not lost on me. I didn't know if it was ironic or appropriate or what. My head was spinning too badly to consider it.

There was one thing I did know, though: I had a partner once, a lifetime ago. We saved each other's lives more than once. And Sasai saved mine today.

"You're right. We do make a good team, Sasai-kun." It might have been the first time I used an honorific with his name.

"That's the head-wound talking."

I laughed. It hurt. "No, I mean it. Let's get out of here."

"We really should get you to the emergency room, Azuma-san. You can walk, right?" Sasai's concern seemed genuine.

I learned at an early age that the world was a dangerous place and that looks can be very deceiving, so you needed to always be on guard. Daily experience reinforced that, but today was different. Today, it seemed like maybe it was okay to let someone in.

"With a little help, I think I can manage."

THE GREY MEN OF GLAMAIG

ANDREW NEIL MACLEOD

"What is that mysterious, conical hill up yonder," said Johnson, "wreathed in sullen clouds like a veritable Olympus?"

"She is called Glamaig, the Red Cuillin herself," said Malcolm their guide, "which means *Greedy Woman* in the English."

"How did it come by such a curious name?" asked Johnson.

"Och, no one can remember anymore."

Malcolm MacLeod had accompanied Boswell, Johnson and their manservant Joseph from Raasay, rowing them over to the Isle of Skye, where they hoped to investigate some of the local myths and legends before tea.

"I should very much like to see it up close," said Johnson.

"I doubt that you will find anyone who would show you the way," said Malcolm. "Glamaig is home to the *Fear Liath Mor*, the Big Grey Man, though they say it is more phantom than man."

"And you have seen this apparition?" asked Johnson.

"Not I, sir," said Malcolm, "But I know a man who has, and lived to tell the tale."

"Then you must introduce us," said Johnson. "I would have an account of the experience for my journal."

"That would be very easy, sir. He frequents the inn where you are lodging tonight."

The four men rode west along the banks of Loch Sligachan, then turned south through the Sligachan Glen, following a winding river of the same name.

Choosing a spot where the current was not so rapid, they stopped to rest by the river, which ran cool and clear over smooth pebbles, sparkling like liquid magic. Joseph tended to the ponies, while Johnson sat on the bank, pulling off his boots and easing his feet into the icy water.

"Tell me," said Boswell, "is it fit for human consumption?"

"The Sligachan is said to grant eternal youth to those who drink from it," explained Malcolm, stooping to fill his flask. "Mrs MacLeod swears by it, and takes a cup each night before she goes to bed."

"And yet judging from those washerwomen," remarked Johnson, indicating with his stick three aged crones washing their linen on the opposite bank, "either they themselves have not partaken, or else your preposterous claim is naught but a fantastic invention."

"Bravo, Doctor!" cried Boswell, glowing with admiration. "Your powers of deduction are remarkable. I stand in awe of your method, by which you sift through the silt of superstition to find the gleaming nuggets of truth within."

"I assure you, Mr Boswell, that it is merely the application of many years of rigorous scientific discipline, and as natural as breathing; once you have the knack, of course."

The three crones glared sullenly at the men on the opposite bank, as Johnson doffed his hat.

After they had rested, the company mounted their garrons and continued on their way, following the course of the river as it meandered through the glen. Malcolm lagged behind Joseph and maintained a sullen silence throughout, while Boswell and Johnson forged ahead, taking note of the strange geological formations, a unique feature of the island, that rose up around them.

The Isle of Skye was, and still is, a land coloured by tales of fairies, fantasms, fauns, Tom-tumblers, melch-dicks, kitty-witches, hobby-lanthorns, Dick-a-Tuesdays, elf-fires, knockers, old-shocks, ouphs, Tom-pokers, tutgots, tantarrabobs, tod-lowries, bogles, swarths, sprites, giants and other outlandish creatures, and as they watched the morning mist roll down from those curiously-formed peaks, it was not difficult to see why.

Finally they came to a rude house in the village of Broadford which acted as an inn; a last stop for drovers before they ferried their cattle across to the mainland.

After a supper of mutton stew with hunks of fresh bread, Boswell and Johnson were introduced to Fergus MacIvor, the old crofter who claimed to have encountered the *Fear Liath Mhor* on the summit of Glamaig.

"It was terrible, sirs," said MacIvor, once they had brought him a whisky to loosen his tongue. "I was coming down from the summit with a barrow of peat for my hearth, when I heard a crunching noise on the gravel behind me. For every few steps I took I heard a crunch, and then another, as if something was coming after me, taking strides five times the length of my own." MacIvor took another sip of whisky to calm his nerves, then continued. "When I looked back into the mist, I saw it. At least twelve feet tall he was, with legs as big as tree trunks, talons like scythes, and grey hair sprouting all over his stinking hide. But the worst of it was the face: The creature's head was shrouded in unearthly light, like a saint from a religious icon. But this was no celestial light, sirs. No, this was diabolical hell-fire! Well I dropped my barrow and ran down the side of the mountain as fast as my legs could carry me. Nothing would induce me to climb Glamaig again for as long as I live. It is eerie up there among the clouds, and there are things that walk the summit that have no right to be amongst the living."

According to those who knew him, Fergus MacIvor was a reliable man. He was also a devout Christian who had never been known to lie, which made his account of the Grey Man of Glamaig all the more interesting.

"Well Boswell," said Johnson, once Fergus had taken his leave. "Now I want to climb that blasted mountain more than ever. What say we rise early and attempt it?"

* * *

When Boswell came down for breakfast his friend was already at table, leafing through his copy of *Essays and Observations*.[1] Boswell wondered why he chose such a ponderous text for a travelling companion.

"Why sir," replied Johnson, glancing up. "If you are to have but one book with you upon a journey, let it be a book of science. When you have read through a book of entertainment, you know it, and it can do no more for you; but a book of science is inexhaustible."

Boswell sat down beside his friend while the landlady brought them each a hearty breakfast of porridge oats. After they had cleaned their bowls the landlady appeared again with a bottle and set it down on the table. "Perhaps ye'd like a touch of something stronger t'set ye up for the day ahead?"

"I am surprised at you, Madam," said Johnson. "Alcohol inflames, confuses and irritates the mind. It is true, as a young man I drank to excess, and suffered greatly as a consequence. But now, in the autumn of my life, I find that abstinence suits my temperament far better."

"Well I don't know about that, sir. But many a Skye man takes his *morning* as we call it, then touches not a drop more. It sets him up for the rigours of the day."

Johnson picked up the bottle and squinted at the label.

"It is called Drambuie, sir. The recipe was given to my grandfather by Bonnie Prince Charlie himself."

"Indeed?" said Johnson. "Well perhaps we may have a taste, after all. What's sauce for the goose is sauce for the gander, eh Boswell?"

[1] Ebenezer McFait, Glasgow University

Boswell was not averse to the notion, and so the landlady poured out a dram for each of her guests. The liquor was strong though not unpleasant, with a honey-sweet aftertaste. Thus fortified, the two men took leave of the inn with a kind word for the landlady, and made their way to the stables, where Joseph had the ponies made ready.

It was easy enough to find their way back to Glamaig alone, following the trail of landmarks Boswell had sketched the previous day. As they trudged across the Sligachan, Glamaig appeared before them, a perfect cone of granite with vast flanks of scree[2], though the peak seemed much taller than it had the day before.

"Are you sure you still want to do this Doctor?" said Boswell.

Johnson was a heavy man already in his sixties, but his youthful spirit rose to any occasion. "To have come so far and admit defeat? I'd be laughed out of the Literary Club."

They rode as close as they could to the base of the mountain and left the ponies with Joseph, promising to be back before nightfall.

It was a steep ascent, but not so sheer they would need the rope Joseph had insisted on. As the two men climbed higher they entered a thick fog which enveloped them entirely. Unable to continue, they decided to sit it out and open the package Mrs Mackinnon the landlady had provided. Johnson unwrapped a pile of oatcakes and a round of crowdie, a cream cheese made with curdled milk and rennet.

"I could get quite used to this," said Johnson, slathering a large chunk of crowdie onto an oatcake with his knife. "The landlady makes it herself, using a secret ingredient handed down from her grandmother. I rose early this morning to discover what she feeds her cows to give the cheese its distinctive flavour. Alas, it is a secret she will take with her to the grave, more's the pity. I could have had Mrs Thrale whip me up a batch when we got back to London."

Once the fog had thinned the two men continued on their way. It was a tedious, laborious ascent, and Johnson suffered the worst of it, panting heavily as he dug his staff into the coarse earth, stopping every few hundred yards to catch his breath.

When they finally reached the summit, they realized that Glamaig was not the perfect cone it appeared to be from the ground, but was joined at the far side to a long ridge of mountains stretching south for many miles.

The views were breathtaking. Using the map provided in their guide,

[2] a mass of small loose stones that form or cover a slope on a mountain

Boswell was able to identify some of the more prominent features. To the south, the long ridge of mountains called simply The Red Hills appeared like the back of some vast, slumbering dragon. Far to the north across the Sound lay the Isle of Raasay, with the distinctive flat-topped peak of Dun-Caan appearing like a tiny hat in the distance. To the west, across a grassy plain dappled with cloud-shadow, the ragged peaks of the Black Cuillin stood in stark isolation.

The two men sat side by side, their backs to a cairn, giddy with elation. The fog had followed them up the mountainside, its formless tendrils enfolding them as they rested. So dense was this fog that they could feel its moisture like a cold breath on their faces. The very air they breathed was thick and soupy. After a while they started to feel sleepy, and before long, both men had fallen into a deep and dreamless slumber.

Boswell awoke with a start. The sun was inching its way towards the horizon, gilding the pink wisps of cloud that lingered over the Black Cuillin. To the east the sky had begun to darken. The evening star shone as remote and lonely as the summit of Glamaig herself.

Boswell roused his slumbering companion. They had slept overlong, and Joseph would be worried. If they didn't leave soon it would be fully dark before they reached base camp.

The fog had receded, presenting an impenetrable wall further down the mountain, like a crown of snow-white hair around a bald pate. It was eerily quiet. Slowly, a looming shape began to make its presence felt in the formless mist.

As they strained their eyes to see, two monstrous shapes emerged. They stood on vast legs, these mountain giants, with elongated arms that hung loosely by their sides. Their heads were relatively small, each crowned by a fantastic nimbus of light; concentric circles of glowing, vibrant colour. Boswell felt his knees weaken in their presence.

Johnson tugged on his friend's sleeve, rousing him to action. *"Run, Boswell!"*

Johnson and Boswell fled south along the ridge of the mountain, staggering blindly in their flight. The fog creatures kept up easily, taking huge strides on their monstrous legs, crunching the ground with each step. Curiously, they did not appear to be coming any closer, but chose to remain within the fog bank, as if this element alone gave them substance.

As the men increased their speed so too did their pursuers, until Johnson finally stopped, resting the palms of his hands on his knees and panting heavily. "It's no use," he gasped between breaths. "Leave me Boswell. I can't go on."

"You must go on!" cried Boswell. "I won't leave you here at the mercy of these ungodly creatures."

Johnson risked a glance at the figures several hundred yards further down the slope. One of them appeared to be mimicking him, bending forward with his hands on his knees. After studying the creatures for a few seconds, Johnson's entire frame began to shake.

"Now, Doctor," said Boswell, "do not weep! We must face our fates like gentlemen."

But Johnson's face was creased with mirth, and his girth was convulsed not with tears, but with laughter. He straightened his back, and did something that took Boswell completely by surprise. He stuck a thumb in each ear and waggled his fingers at the giants, sticking out his tongue like a naughty schoolboy. To Boswell's amazement, one of the creatures did the same. Johnson let out a wild cry of triumph, and snatching the hat from his head he waved it in the air, only for his actions to be mimicked again by one of the fog creatures.

"Ah, Boswell, fear will make fools of us all!" cried Johnson. "Those creatures nearly unmanned me, and yet there is a simple enough explanation, had we the wit to look. Follow me if you dare!" Johnson strode forward, with a reluctant Boswell close behind. As they entered the mist the creatures seemed to evaporate into thin air. "See, Boswell? They are only shadows after all!"

Boswell struggled to keep up with his friend. "But shadows that stand vertically twelve feet tall, with glowing lights on their heads?"

"Aye, it is a singular phenomenon, I'll grant you, but let's get ourselves safely down the mountain first, and I'll explain later. Joseph must be frantic with worry."

They were not out of the dark yet. It was a torturous descent, and the mountainside had many hidden dangers, the threat of a landslide not the least of them. Midway Johnson lost his friend in the mist and had to locate him by sound alone. After that, they found that Joseph's rope came in handy after all, and used it to tether themselves together.

* * *

Joseph wept openly when he saw the two men staggering down the scree and ran to embrace them, a gesture which they found both strangely touching and vaguely uncomfortable.

Happy to be reunited, the three men clambered onto their ponies and began the long trek back to the inn. Though Boswell had a thousand nagging questions, Johnson remained tight-lipped, collecting his thoughts until he was comfortably ensconced, and could hold forth in his own inimitable style.

When finally the three men were seated round the table with a supper of cold ham and a tot of Drambuie before them, Johnson was ready to begin.

"Ebenezer MacFait first describes the Brocken Spectre in this book, under a chapter entitled *'Some Phaenomena Observable in Foggy Weather'*," he said, placing his well-thumbed copy on the table.

"The... Brocken Spectre?"

"Yes. Named after a mountain in Germany, the Brocken, where the phenomenon was first recorded. Put quite simply, when the sun is at a certain angle behind the subject, it casts a shadow vertically against a fog bank. It only occurs at dawn or dusk, when the angle of the sun is in a direct line with the slope of the mountain. The figure appears gigantic, because there is no frame of reference in the fog."

"I see," said Boswell. "But how do you explain the corona of light which appeared around the creature's head?"

"I am glad you asked that," said Johnson. "Have you never observed a rainbow?"

"Yes, of course," Boswell replied.

"Well it follows the same principle. Rings of colored light appear when the sun's rays illuminate a cloud of water droplets, creating the eerie halo effect, often referred to as the *glory*, which so unnerved us on the mountain. To quote MacFait: *'All shadows converge towards the anti-solar point, where the glory also shines.'* "

"And yet I *heard* those creatures," argued Boswell, "or at least, I heard their footsteps. The last I heard, shadows have no material substance with which to affect the world around them."

"Quite right, Boswell, and I must confess, the phenomenon threw me at first. But you will have observed that the summit of Glamaig is littered with countless broken shards of rock?"

"Yes, but what of it, Doctor?"

"As fog thickens and thins," explained Johnson, "the temperature fluctuates, and rocks expand and contract, causing them to shatter. When this happens on a slope, the smaller pieces of rock will tumble. These actions are, in fact, entirely responsible for the scree of which Glamaig consists. Even on a calm day, rocks will shatter and slide down the mountainside, which may account for the sounds we perceived, in our heightened state of alarm, as footsteps."

"Remarkable, Doctor. But how do you account for MacIvor's lurid description of the Grey Man's physical appearance, the fur and the claws, where we saw only formless shadows?"

Johnson forked the last piece of cold ham into his mouth and chewed thoughtfully. "Who knows what the mind, alone in lonely places, can create out of nowhere," he said, wiping his mouth with his napkin. "Illusions and hallucinations may be brought on by isolation, or from sheer exhaustion alone; indeed, the wind itself may whisper strange things to a man, when alone on the high places of the earth. All that was needed was the phenomenon of the Brocken Spectre, remarkable enough in itself, for MacIvor to fill in the details with his own vivid imagination. Shadows, it would seem, are as a blank canvas to the suggestible mind."

"Well I must congratulate you, Doctor Johnson. Your logic is water-tight," said Boswell. "Truly, your mind is a lantern, dispelling the gloom of ignorance wherever you go."

"And yet," said Johnson with feeling, "there were moments up there, Boswell, when I felt a closeness to some ineffable presence, a mystery far greater than ourselves, which inspired in me the profoundest sense of reverence and awe."

"Quite so, Doctor Johnson," replied Boswell. "It is not for nothing that Scotland is known as *God's own country.*"

"Well I wouldn't quite go that far," said Johnson. Then turning to the landlady he said, "Madam. Would you be so kind as to bring me some of your delicious crowdie? And heavens to goodness, don't scrimp on the oatcakes this time."

The first in a new semi-regular collumn on gaming within the occult detective sub-genre.

THE CLUETRAIL CONUNDRUM

PAUL STJOHN MACKINTOSH

It won't have escaped many fans of occult investigation that an occult investigation needs investigation as well as occultism. Some tales of occult investigation may immediately proceed to the solution through a seance or a spontaneous revelation, but I bet they're few and far between, and probably not among the best in the genre. For the writer of investigative fiction, occult or otherwise, much of the challenge (and pleasure) consists in constructing an interesting, credible, and engaging investigation. And the same applies to the creation of investigative roleplaying games.

Clues – finding, interpreting, solving them – are actually one of the most problematic and sometimes controversial parts of investigative roleplaying games. One issue comes at the adventure design level. The problem is the same as if you insert some kind of number puzzle, or anagram, or other conundrum requiring mental effort, into a computer game. What about those players who can't solve it? What if that puzzle has to be solved before you can proceed to the next stage or level of the game? Won't you be doomed to frustrate and infuriate at least some of your players if you do put such a puzzle into your game?

Tabletop pen-and-paper roleplaying games are usually a good deal more flexible and less linear than computer games, and allow for far more solutions to specific situations. All the same, some adventures can fail if you don't provide alternative workarounds to particular puzzles, or leave the solutions where they can be found by different means. You can't always expect your players to be able to solve particular puzzles – nor should you. After all, characters in roleplaying games are supposed to simulate particular skills and capabilities that their players don't necessarily have. A mediocre player can create a genius detective in the same way they can create a gifted fencer, just by focusing on the requisite attributes and skills. And if the player can't solve a specific clue by themselves, they can (and probably should) get the answer by rolling against the character's Intelligence.

This points to a more fundamental problem than just good adventure design, though. The single point of failure in roleplaying games can happen on the same basis as a badly thought out computer game, where one wrong turn or missing clue can simply stall the adventure forever. And this is controversial because a lot of gamers insist that it never happens.

The problem comes when players fail their Spot Hidden rolls (or whatever the equivalent is in their gaming system) to find clues. The clues are essential to progress the game. The dice are there to mimic effort, and to provide an exciting and challenging random element to the game. But if the dice dictate that you don't find a particular clue, what happens to your progress in the game? As Kenneth Hite explained in his introduction to *Trail of Cthulhu*, the game's Gumshoe system exists "to solve a problem that many people found with running *Call of Cthulhu* – one bad die roll can derail an adventure."

This has become such a cliché in gaming circles – and across gaming systems – that it's been encapsulated as the "Spot Hidden Fail" (a.k.a. The Total Scenario Kill). I've had countless direct, personal experiences told to me by people who had exactly this happen. I've read of countless more examples. I've also had just as many rebuttals and dismissals from gamers who have sworn that their favourite system can't produce such glitches, that the players must be at fault, that the problem is purely poor adventure design, etc., etc., ad nauseam.

I won't deny that bad adventure design can throw up such issues. I won't deny either that players can be incredibly obtuse, and miss what's right in front of their (character's) eyes, just as readers can miss the most obvious clues in an investigative story. But I have seen a strong strain of defensiveness and denial in the gaming community about this issue – to the point that people will deny or ignore direct accounts of it happening, right in front of their eyes.

The last time I brought up this issue in a major online roleplaying forum, most of the discussion revolved around *Call of Cthulhu*, which is practically the originator and pioneer of investigative roleplaying – as well as occult investigation in RPGs, if you class Lovecraftian horror as occult-ish (Madame Blavatsky it ain't). The debate was about how far gamemasters should be free to interpret RPG rules as written and come up with workarounds where the rules themselves created problems or stalled the game – and the main case in point was the Spot Hidden Fail. In the course of that discussion, after ignoring numerous examples of Spot Hidden Fails thrown up in the debate, one *Call of Cthulhu* veteran started a separate debate where he attempted to prove that almost no adventure written for *Call of Cthulhu* ever had a potential Spot Hidden Fail built in. And that game has had a lot of adventures written for it.

My position is that, dice being what they are and human nature being what it is, if something can go wrong, it will go wrong. If you structure an investigation around a trail of clues, then at some point, someone will always lose the trail. That can apply in investigative fiction, of course, as well as in investigative RPGs. So if you accept that the Spot Hidden Fail exists as a problem, rather than attempting to dismiss it as some kind of RPG community urban legend, how can you foolproof your adventures and your RPG rule systems (as far as you can) against it?

One approach, as Chaosium itself recommends in the latest editions of *Call of Cthulhu*, is simply to put the key clues in plain sight where no one needs to roll to spot them. That may lose a little drama but it certainly ensures that no one ever gets lost for want of clues.

Another, as devised by veteran game designers Robin D. Laws and Kenneth Hite, is to make all clues instantly accessible without a roll. As exemplified in the Gumshoe system, this approach makes certain specialist abilities 100% successful – use it and you immediately get your clue, because that's what you need to move forward.

A third, as used in *Call of Cthulhu* offshoot *Delta Green*, is to make certain skills automatically successful at certain levels of expertise. If you're a 60% Archaeology PhD, you automatically unearth your clue, no roll required.

A fourth, as used in the Powered by the Apocalypse family of games and elsewhere, is failing forward. If you roll and fail, you get your clue, and something interesting happens. It may be nasty interesting, but at least it won't crash the game, right?

Yet another, developed by gamer Justin Alexander, is the Three Clue Rule: "For any conclusion you want the PCs to make, include at least three clues." This doesn't just apply to finding clues, but also interpreting them – because after all, any clue is only as good as the deductions made from it.

And thereby hangs a tale, because astute readers will have realized that finding the clues is only half the problem at best – and probably less. As Kenneth Hite stated in *Trail of Cthulhu*, "investigative scenarios are not about finding clues. They are about interpreting the clues you do find." This is a topic that may well be relevant to occult investigative fiction as well as gaming, and certainly has parallels in forensics and intelligence. Understanding the clues you find is often the real challenge – making sense of a deluge of raw information and trying to sift out the common threads that underlie the data. Pattern recognition is surely at least as important a skill in all these areas than the simple ability to unearth yet more details.

Yet the emphasis in RPG design still seems to go on unearthing clues rather than interpreting them. As detailed above, the RPG community has come up with more than enough answers to the Spot Hidden Fail. Yet Gumshoe adventures, to cite only one system, are still constructed on the basis of clue trails, where one clue, or at best a few, leads to the next scene in the adventure, and so on till the end. Perhaps the mechanics of game design shouldn't be focused there at all. After all, the most dramatic scene in a detective story is usually the final denouement, where the investigator either realises or reveals the overarching interpretative plot that explains and unites all the disparate clues. Perhaps that's what the mechanics should be structured to deliver. But that's probably another subject for another time.

VINNIE DE SOTH AND THE SAUCER PEOPLE

I.A. WATSON

"Actually," Vinnie de Soth confessed, "I have never been probed."

Moomin looked sceptically at the young man in the scruffy jeans and faded Clannad T-shirt. "What, never?"

The sky-watcher beside Vinnie was music-festival grungy but not unattractive in her yellow anorak and hiking boots. One half of her hair was sprayed pink and the rest was green. The seeker's name really was Moomin, given to her in a New Age birthing ceremony at Iona during a juxtaposition of Venus and Mars in the late 90s, by a confused mother who had either never read Tove Jansson's translated *Mumintroll* children's books[1] or else had mixed up the names and thought Moomin was a Tolkien character.

"I try to avoid being abducted, generally," Vinnie explained. "By aliens or anybody else. Especially family."

"It's not abduction," Mr Testable insisted. Fussy and fifty, he sipped weak Oxo broth from the plastic cup top of his thermos flask.[2] "It's a blessing of revelation."

"Like hell it is!" objected a middle-aged fat American who sat encircled by detection technology: cameras, vibration sensors, spectrometers, Geiger counter, orgone detector, and much more. "It's prelude to invasion, buddy. It's state-sponsored psychoterrorism, the co-option of the collective unconscious by capitalist meme-waves – the colonisation of our underthoughts. That and tubes up our asses."

Moomin objected. "It's not always like that. You're making us sound like idiots. You're being reductive about an unfathomable psychic experience."

Across the cleared crop space was a bearded academic in elbow-patched tweeds, squatted on a camping stool with tape recorder and binoculars. "No

---

[1] The hippopotamus-like fairytale Moomins appear in nine books and a comic strip by Finnish illustrator Tove Jansson. They have starred in several television series, films, and a Moomin World theme park in Naantali, Finland.

[2] American readers may know this helpful camping accessory as a vacuum flask or a Dewar bottle.

experience is unfathomable," he corrected Moomin. "Otherwise why are we out here?"

"You expect to encounter aliens?" Vinnie asked him.

He shook his head and handed over a business card that proclaimed him *Dr Alwyn Shore, Lecturer in Relative Psychologies* at one of the provincial universities. "I expect to encounter people who expect to encounter aliens," he explained. "What I want to know is, if enough people come together and want to see something, whether something will appear."

"Jungian rubbish," muttered the American. "The Psychosocial Hypothesis is just a way for believers who are too ashamed to admit they believe to explain their interest."

Vinnie was aware that pioneer analytical psychologist Karl Gustav Jung had published *Flying Saucers: A Modern Myth of Things Seen in the Skies* in 1958, and that it had eventually spawned a strong body of support for the idea that 'UFO' sightings beyond those caused by weather balloons, misperceptions of planetary bodies, and marsh gas were a psychological rather than astronomical phenomenon. A good way of starting a fistfight amongst UFO investigators was to stir up the Psychosocial Hypothesis dedicats versus the Extraterrestrial Hypothesis supporters.

"So you're here to see if these folks can *believe* a flying saucer into appearing," the jobbing occultist checked.

"We can't believe it into being, but we can shift our consciousness so we can see it," Moomin argued. The American snorted and Dr Shore made a note.

"What are you, then?" Mr Testable challenged Vinnie. "Saucerhead? Crystal-worshipper? Corndancer? Fourth Ager?" Mr Testable was proudly a Nuts-and-Bolts Evidencer, and if called out on the matter could produce the small plastic bags with extracted implants to prove it. He had also offered to demonstrate his sachets of irradiated soil, but Vinnie had declined.

"I'm here with an open mind," Vinnie promised the ring of people waiting in the crop circle by night under the starry Avebury sky.

"You *think* you're here with an open mind," the American with the advanced alien-surveying equipment warned. "You're actually responding to high-frequency radio algorithms that are programming you *how* to think." He demonstrated some EMF readings that proved it.

"Everyone is here for a different reason," Moomin summarised, "but we all came. Isn't that lovely?"

Vinnie nodded absently, but actually he was looking about for the reason he had come. It was the night of the 1st of May, the ancient festival variously called Cétshamhain, Clan Mai, Beltane, or the Feast of St Joseph the Worker, a time of seasonal mystical significance. The gathering was up beside the bank of the Ridgeway, Britain's longest and oldest prehistoric trackway.[3] But most significantly, it was centred upon the strange circle in the corn that had appeared three nights earlier.

"Everybody is looking for something," Vinnie agreed with Moomin. "I'm looking for a woman called Carol Chanter. Do you know her?"

Caroline Chanter had disappeared after a vigil in the crop circle the night before last. Carol's mother was worried about her. When the police investigation had stalled she had called in a jobbing occultist whose humble one-line Yellow Pages entry had caught her eye.

Moomin frowned. "Bit older than me? Wears Doctor Who scarves?"

That fitted the description offered by Mrs Chanter, but Vinnie proffered a Polaroid photo of the missing person snapped at the memorial marker atop the Uffington Castle Iron Age hillfort.

Moomin unwrapped a mint and popped it into her mouth as she studied the picture. She thoughtfully passed one to Vinnie too, and encouraged him to take another for later.

"You recognise her?" the jobbing occultist checked.

"Yeah. I thought that was her. She posts online as CallMeOsgood7.[4] It was her post about the Goreston Farm circle phenomenon that alerted us all."

In addition to Moomin, Mr Testable, Dr Shore, and the well-equipped American who sat with Vinnie in the flattened-down thirty-three foot corn ring there were a score of other hopeful alien-watchers and would-be contactees lurking nearby. Some were up on the Ridgeway itself, others circled a traditional 'balefire' a little way down the slope, and a few separated out alone in hopes of exposure to the uncanny. One very hairy adherent of crystal culture was sat meditating inside a complex pattern of

[3] The Ridgeway or Iknield Way is around 5,000 years old and traverses Dorset to the Wash; some argue for its continuation into Lincolnshire. It passes near many Neolithic, Iron Age and Bronze Age sites including Avebury Stone Circle, the Iron and Bronze Age hill forts of Barbury Castle, Liddington Castle, Uffington Castle, Segsbury Castle, Pulpit Hill and Ivinghoe Beacon Hill, the Neolithic chieftain burial tomb Wayland's Smithy, the Uffington White Horse chalk carving, and the five-mile Iron Age earthwork of Grim's Ditch. It also passes the private drive of Chequers, the Prime Minister's country retreat.

[4] Presumably derived from Petronella Osgood, an intrepid *Doctor Who* recurring character noted for her Doctor cosplay.

quartz, but his lattice was so ill-aligned that Vinnie had gone over there and corrected it just to avoid the headache he'd have got otherwise from the screaming selenite and angelite.

"When did Carol tell people about the place?" Vinnie wondered to Moomin. "How did she know about it?"

"The farmer tipped her off. She posted about it the morning after it appeared."

"This would be the farmer who's charging twenty quid to everybody who wants to sit up here overnight and sixty quid for camping?" In these straitened days for agricultural endeavours it was not unknown for enterprising landowners to arrange for strange crop-pattern phenomena as a way to enhance their incomes.

"It's only fair. Extraterrestrial forces destroyed his livelihood."

"Well, that or a few blokes from the pub with ropes and planks."

Moomin scowled. "You're a sceptic?"

"Not really. I believe in all kinds of weird gubbins. But I also believe in blokes from the pub with ropes and planks."

Moomin looked around the flattened circle of cereal. "That sort of trickery does go on *sometimes*," she admitted reluctantly. "But look at this evidence. None of the stalks of corn are broken, just pressed down. That's a sure sign of an earth vortex."

"Earth vortex," Dr Shore repeated, making another note.

"There's no such thing as an 'earth vortex'," Mr Testable tutted, more in regret than anger. He set down his Oxo. "What you're seeing here is a side effect of a gravity wave from a hovering saucer. Judging by my readings I'd say it was a Type 3, maybe 3a. That's why it's worth keeping watch to see if there's a recurrent encounter."

"Bull*crap*!" the American interrupted crossly "You'll never discover anything of worth with that literalist attitude and those ridiculous classifications," he warned the Nuts-and-Bolts adherent. "It's well established that most UFOs are plasma waveforms projected over cosmic distances, perhaps over dimensions, and controlled by world governments to enslave their populations."

"Or," Moomin offer brightly, "what we're seeing here, actually sitting in, is an encoded message to humankind if we can only understand it. It could be a test, a means of determining whether we are spiritually advanced enough for full contact."

Vinnie had checked the site for occult activity. The whole area around Avebury was awash with leys, making it hard to distinguish any actual magic over the background noise. The placing of the circle was possibly significant, but he couldn't find any evidence of ritual or spellcrafting. The half-dozen young people around the Beltane campfire down the hill were generating more accidental arcane energy with their dancing – and some covert liaisons in the cornfield – than whatever had made the corn circle had produced.

"I'm not sure about this site, that's all," Vinnie told them. "But a woman came up here two nights ago to keep vigil and she's not been seen since."

"Taken," Mr Testable concluded. "Snatched. Fourth-kind contact."

"If she did pass through a chymeric window then she might endure a lost-time experience," allowed the American. He sounded envious.

"Or she could just have hooked up with any of the seekers who visit these sites for, um, other kinds of close encounters and gone off with him or her," Dr Shore pointed out.

"Her mother's worried about her," Vinnie explained. "But the police aren't taking the case too seriously. Carol was an adult with every right to wander off and not return home if she wanted. They made a search of the area, asked a few questions, and more or less decided she'd probably scooted off the London or somewhere."

It was quite possible that CallMeOsgood7 had indeed decided to chuck in her shelf-stacking job at the local Sainsbury's supermarket and seek her fortune elsewhere, but the jobbing occultist suspected that she would probably have taken her DVD collection with her. And kept up her blog.

Moomin shuddered and looked around at the darkened cornfield. "If she was abducted, it was from here."

That was probably true whether the kidnapper was mortal or alien.

"What about last night?" Vinnie wondered. "The night after the circle got made, it was pretty much just Carol up here. Nobody else has admitted to being around then. Tonight we have a busload of would-be contactees and investigators. What about yesterday?"

"I was here," Mr Testable testified. "I was the first on site yesterday, camped down at the farm, up here taking readings: background radiation fluctuations, ground resistivity, temperature hotspots, EM activity; the usual. No point trying to make proper scientific measurements once the amateurs and saucer-chasers turn up. They just end up taking photos of ghost-orbs, misidentifying Venus, and worshipping the 2343 flight from Heathrow to Dublin."

"Whereas you use science," Dr Shore scorned the amateur.

"Sure we do," the American defended his misguided Nuts-and-Bolts colleague. "After all, the Nation-States use science against us."

"So cynical," Moomin objected. "No wonder we don't receive contact. We're all too busy quarrelling."

"Did anyone quarrel with Carol Chanter before she vanished?" Vinnie asked. He showed the photo round again. "Moomin has met her before. What about the rest of you?"

Mr Testable had encountered the young woman previously on occasion but had not spoken to her. The American who still preferred anonymity had only arrived in the UK yesterday, having flown in to inspect 'the Goreston Farm event.' Dr Shore denied any knowledge of the woman in person; he had seen her blog report but had not known her real name until now. A couple of the outlier investigators recognised Carol too but had little useful to add other than that "she was around sometimes".

Vinnie didn't ask about the material in Carol's diary, some of which had filtered into her online reports, but Moomin had evidently read some of CallMeOsgood7's blog comments. "She was wrong about the origin of UAPs, but she had some interesting data. She'd been searching around the Goreston Farm area for weeks before the cereal event."

Vinnie recognised UAP as 'Unidentified Aerial Phenomenon', the preferred modern term which had replaced the newspaper cliché 'Unidentified Flying Object' on the grounds that (1) it might not be an object, (2) it might not be flying, and (3) it got serious researchers laughed at less – somewhat.

"The very name Goreston Farm is significant," Moomin went on. "A gore-stone was presumably a place of ancient blood sacrifice, a soft place between one world and another."

Vinnie refrained from pointing out that *gore* was an ancient term for a triangular piece of land. Or it was Middle English for dung; the modern adjective *gory* came from that, meaning filthy, and it hadn't been coined until the 14th century, three hundred years after 'Gorstan' had been mentioned in the Domesday Book. It wasn't as exciting a derivation as a young mystery-seeker might hope.

"Carol had mapped some ley lines," Moomin further reported. "Not just physical alignments of natural and Neolithic features, but energy routed in serpent-tracks."

"Not *that* again," Mr Testable objected. "One cannot dowse for alien craft."

"Not with branches and twigs and crystals on chains," the American agreed, "but if you use high-end magnetometers and ground radar to track small waters..."

"It's possible to see them just by psychic questing if only you…" Moomin insisted.

"There are all kinds of natural explanations for the dowsing phenomenon," Dr Shore pointed out. "There's an inherent self-bias that…"

Nobody was listening. Everybody had a point of view to espouse.

"One at a time," Vinnie pleaded. "Moomin, you were saying what Carol had written about?"

"She'd been mapping ley lines," she repeated, glaring challenge at the others in the corn circle to contradict her. Vinnie, who had been taught that *ley* was an Old English word for line, did not point out the tautology.

"What did she think she'd found?" enquired Dr Shore.

"She thought that the leys were realigning. That happens sometimes, when big electric pylons are planted in their way or old stones get bulldozed, but there didn't seem any explanation for this change. Unless it was that deep drain digging at Brockhampton."

"That's why she was already in the area when the farmer had his lucrative cereal phenomenon?" Vinnie reasoned.

"I guess so. And why she was first on the spot, on her own that first night."

Vinnie frowned. What he really wanted now was a quiet room and a pocket calculator to do some arcane mathematics. There were seasonal and astrological factors, locational data, ley patterns, belief vectors, and possibly even accidental ritual triggers, and he wanted them laid out in a matrix where he could try to derive some model about what, if anything, had happened here. But refuge and calculator were denied to him.

"What was that?" the American interrupted urgently. He checked his instruments. "34 degrees from north, about 60 degrees from horizon."

"I saw nothing," Mr Testable scorned, but he trained his field glasses in that area and swept the sky.

"There was a bright light, quite transient. But my video cameras weren't pointing that way."

There was a second flash. This time everyone in the circle saw it.

Mr Testable checked his phone. "There are no scheduled flights in that area. It might be a night-glider or lantern balloon, but it was momentary, not continuous."

"There it is again!" Moomin cried out, pointing.

Vinnie glanced down the hill. The balefire dancers had all vanished into the corn now, but he could just glimpse the crystal specialist on the Ridgeway and a couple of other solitary watchers. None of them seemed to have spotted the flash.

Dr Shore watched the American reposition his equipment in the direction of the light. "What do you think is happening right now?" he asked the others.

"Too early for determination," Mr Testable told him brusquely. "We're not idiots, you know, misidentifying every little thing as aliens. You'd like to caricature us that way, but we are probably better trained to discern natural causes and aircraft activity than any 'expert witness' you care to produce."

"We're perceiving something," the American reported. "Collective perception, too. Even you, Shore. There it is again. Looks nearer."

"Do *you* think it's aliens?" Moomin asked Vinnie hopefully.

"Not if they're coming to probe me," the jobbing occultist answered. "Honestly, I don't know if aliens even exist. Lots of weird things do, but I've never met an alien. That I know of."

"Then how do you explain the thousands of encounter cases that have been recorded and investigated?" Mr Testable challenged.

"Ninety-nine percent were misperceptions or fraud," Dr Shore replied. "The others were genuinely perceived but were something from within the human psyche, not from Zeta Reticuli."

Vinnie actually suspected fairies. They'd been in the business of leading humans along with strange lights and noises, causing lost time experiences, and occasionally kidnapping and having sex with mortals long before pilot Kenneth Arnold's sighting of nine shiny unidentified flying objects flying past Mount Rainier at speeds in excess of 1,200 miles an hour on 24th June 1947. And fairies loved to dress up and play new games.

Of course, Mr Testable might argue that those encounters were aliens too, just explained by the beliefs of past eras. Then again, he'd not visited the Many Coloured Lands since childhood or attended the Seelie and Unseelie Courts, whereas Vinnie had survived an unusual upbringing and had once killed a Redcap with fridge magnets.[5]

The jobbing occultist triggered his arcane sight, overlaying his perceptions of the magical energies around him on top of his observations of the increasingly-excited watchers. He had to revoke his spell immediately; the swirl of twisting forces thrumming across the circle was blinding.

But natural, not invoked. Was it just Beltane near Avebury and Silbury Hill, or something else?

[5] *'Vinnie de Soth and the Harvest Feast'* in *Vinnie De Soth, Jobbing Occultist* (2015), Chillwater Press ISBN 9798772386804

"It *is* getting closer!" Moomin called. "And flashing more frequently."

"It's not appearing on my screen," the American complained. Then he swore. His camera battery had failed.

"Classic saucer-encounter behaviour," Mr Testable lectured him. "That's why such equipment is useless, except for telling you when there is rogue electromagnetic activity. Look at my magnetic compass, spinning like a top!"

"That can be easily faked," Dr Shore insisted, but he sounded nervous.

"It could all be mass hysteria and shared delusion," Vinnie cheered him up. "The madness of crowds and all that."

"They're coming!" Moomin celebrated, jumping up and down and waving at the skies.

One by one, each of the American's remaining gadgets failed.

Outside the broken-down-corn ring the other observers seemed to have noticed nothing. They continued their vigil ignorant of the pulsing light that closed ever nearer to them.

"Perhaps we're the ones looking in the wrong direction," Vinnie muttered to himself. When something wanted him to look up, his natural inclination was to check his feet.

He surreptitiously crafted a more sophisticated occult detection that included filters much like sunglasses. He tried to make sense of the urgent tangle of energies that swirled around the perimeter of the circle. There was plenty of natural force there, but twisted in ways he could only describe as *unhealthy*. Strands extruded though the ground like feelers, clustering beneath each of the close encounter percipients.

"It's showing us things," the jobbing occultist growled.

Moomin misunderstood him. "It will," she promised. "Oh, to encounter the infinite... the divine!"

"Yeah, that would be nice," Vinnie accorded; but the fierce throbbing energies that surrounded and suffused the crop circle did not feel very holy. They felt angry and malicious and... hungry.

If Caroline Chanter had encountered something like this two nights ago then she hadn't run off to London, unless she'd fled there in panic. Carol had probably never had the chance to escape.

The light in the sky was almost strobing now. Its flashes were tinged with red, followed by a brief crackling echo The American fumbled in his equipment bag for an old fashioned film-fed optical camera, something not dependent on batteries. Dr Shore shook his head in disbelief as the UAP closed on the corn circle.

"This may well become a significant close encounter," Mr Testable said in his fussy way. "An encounter of the fourth or fifth kind... or even the seventh.[6] If necessary, Miss Moomin will have to undergo sexual relations with our visitors."

"*You* screw 'em!" the young skywatcher told her colleague. She looked nervously across at Vinnie. "Do you think we should get out of here? I mean, the unknown is beautiful and that, but..."

"You do not like the way that phenomenon feels?" Dr Shore suggested.

"Not one bit," Vinnie agreed with Moomin. "That light-'n'-sound-whatever-it-is does not feel to be bringing a Message of Peace For All Mankind."

"It's almost upon us," Mr Testable pointed out. "It is too late to retreat."

Vinnie had to agree. He alone could perceive the vicious snarl of whirling energies crackling about the perimeter of the corn ring, and how they were beginning to discharge lightning-crackle tendrils between the globe above and the circle below.

"Time for some camping," the jobbing occultist decided.

"For what?" Moomin asked, trying not to panic. It was one thing to be an ambassador for a first contact, but being fried or possessed by some unknown alien force didn't seem such a great idea.

The American had another question for the scruffy Clannad fan. "Why are you sticking tent-pegs into the ground?"

Vinnie ranged out a ten-foot diameter circle of metal spikes. Moomin's eyes widened with understanding. "Iron... cold iron, it's magnetic. There are reports that it interferes with ley lines... channels them, even."

"Not quite," the jobbing occultist instructed. "Not until you use this spool of thin copper wire and connect all the pegs together. Then you have..."

"A sort of Faraday cage?" the American guessed. "An insulated patch in the middle of whatever's happening!"

Vinnie added a twist of magic to the arrangement. Suddenly the blaring arcane energies swirled around the tent-peg array – but not through it.

"Get inside this ring," the young occultist warned the watchers. "Now!"

[6] UFO researcher J. Allen Hynek's *The UFO Experience: A Scientific Inquiry* (1970) proposed a classification system for UFO experiences. Close Encounters of the First Kind were sightings of UFOs; Second Kind left physical evidence; Third Kind involved sightings of alien beings. Since then, the 'Hynek Scale' has been expanded to include Fourth Kind, abduction, Fifth Kind, direct communication with aliens, Sixth Kind, death of a human or animal, and Seventh Kind, sexual reproduction.

Moomin leaped into the insulated circle. She gasped as she entered the calmed area; she hadn't realised how oppressive the atmosphere had got under that pulsing luridly-red sky-light.

The American abandoned his bag of kit and joined Vinnie and Moomin. After a moment's hesitation, Mr Testable did the same.

"Really?" scorned Dr Shore. "You're going to submit to this mass delusion? Bow to superstition?"

"You *don't* want to stay out there," Moomin urged the psychologist. "You can write about how dumb we are later. Just get inside this ring now. Please!"

The university man shrugged and allowed himself to be coaxed across the copper wire. Vinnie checked the arcane helixes blistering around the perimeter of his sanctuary. The diverted telluric energies were becoming toxic now.

The cornfield swayed as strange currents played over it. Other sky-watchers beyond the cereal circle began to apprehend something. The amorous Beltane celebrants became nervous in their corn bowers. The hirsute crystal-seer was disturbed from a trance when each of his vision-stones fractured in turn; his focus serpentine shattered as if struck by a hammer. He scrambled away, unaccountably terrified, and hared after the half-clad balefire-dancers towards the car park field.

"This may not be a benevolent encounter!" Mr Testable cried over the screaming static of the overhead phenomena. "This may be a Type 5 or 5a visitation. You said that Carol woman was abducted?"

"It's always probing," Moomin warned him, "or impregnation. My friend Zoe had a close encounter at Glastonbury when she was on LSD and she was impregnated."

Vinnie thought that that was a common problem at music concerts, and aliens were not required. Or it might have been Beeping Ron, who notoriously dressed as a visitor from another planet to seduce gullible festival-goers using nothing more than a sink plunger, an egg whisk, and a voice modulator. In any case, it wasn't helpful to worry about it *now*.

"This will make a fascinating article," Dr Shore celebrated. He had abandoned his now-useless audio recorder and had returned to his pocket notepad.

"The wires are protecting us," the American judged. "They're probably diverting the phased microwave pulses being beamed at our brains by whatever agency is trying to keep us from discovering the truth."

Vinnie was unhappy about the levels of arcane force sloshing about the site right now, easily in the terrathaum range, the psychic equivalent of a category 4 hurricane. "Could you all stop believing things, just for a bit?" he asked irritably. "Could you stop believing them into existence?"

Mr Testable didn't pay any attention. He dropped to his knees and said, "They're coming."

The malevolent red light throbbed faster, and then a white beam cut through it like a searchlight. Something in that actinic glare moved, resolving into three tall impossibly-thin figures.

"Do you see them?" Moomin pleaded with Vinnie and the others. "Tell me you see them."

"We see them," the American agreed. "The microwaves are penetrating our defence barrier."

"We do appear to be perceiving three entities," Dr Shore admitted.

"They're too thin to be Nordics," Mr Testable mumbled unhappily; reports of the tall, blonde, Aryan-like Space Brothers, variously from the Pleiades, Venus, or Agartha, generally portrayed them as benevolent to mankind. "Too tall to be Greys. More like the Flatwoods Monster.[7] Can you see scales? If so it could be Reptiloids."

"Like your Queen?"[8] worried the American.

"Do they have penises?" worried Moomin.

The three elongated creatures glided forward, gelid white shells glowing internally, flecked with pale red veins. They stretched out arms to beckon the watchers from Vinnie's ring.

"Don't go," Moomin said suddenly. "They're *hungry.*"

As soon as she said it, everyone else felt it too, a rapacious appetite that might never be satisfied, a yawning void of devouring.

"This... a new classification..." Mr Testable whimpered.

"Psychic vampires," the American gasped. "Mind-sucking parasite waveforms..."

Carol Chanter hadn't had a wire ring to shelter in, to protect her from

[7] The Flatwoods Monster, a.k.a. the Braxton County Monster or the Phantom of Flatwoods, was reportedly sighted after a bright light in the sky over Braxton County, West Virginia, United States, on September 12th, 1952. It was described as a "man-like figure with a round, red face surrounded by a pointed, hood-like shape."

[8] A conspiracy theory popularised by former broadcaster David Icke holds that shape-shifting reptilian aliens control the planet by taking on human form and using political power to manipulate human society. The British royal family are amongst those who are disguised reptiles dominating the Earth.

the insistent pull of the three creatures. Had she been lost to their endless hunger?

"What do you think, Dr Shore?" Vinnie asked the group's psychologist.

"I'm not sure," he replied. "Perhaps one of us should go out and greet them?"

"We can't go out there!" Moomin cried. "It would be the end of us."

"A storm," Vinnie told himself. "A perfect storm."

The American frowned. "What? What do you mean?"

"The disturbances of those drain-works that Moomin mentioned, disrupting the ley-grid. The season. Some farmer deciding on a corn-circle attraction. A long-neglected mostly-broken telluric-arcana routing system that hasn't been maintained since the Druids. Believers bringing their beliefs into a high-malleability environment. Carol, sitting up here like an ancient sacrifice."

"But that was two nights ago," Moomin protested. "This is Beltane!"

"And Carol was the entrée. Tonight we have all the old dominoes lining up, along with a rather spontaneous Beltane fertility ritual down there in the field and a whole bunch of volunteers looking to experience the uncanny. If you pick up the phone and random-dial, chances are that *somebody* will answer."

"Like that?" Dr Shore suggested, gesturing to the luminous beings who had now spread out to ring the tent-peg circle.

"Everybody was calling out and the line was connected. But who connected it?"

"We've probably been set up," the American agreed. "CIA, or DARPA, or Bilderberg."

"Is this invasion?" Mr Testable whimpered. "Steven Hawking warned us... he *warned* us!"[9]

Vinnie couldn't read much outside his protective wire; he'd insulated the barrier against magic. But he was ready to dismiss his previous theory about the fey. This didn't smell of Faerie. There was none of the mischievous glee that accompanied even the most malevolent of fée encounters. There was only malice and terror, and a yawning need to be fed.

He considered the possibility of actual aliens. If such things existed then presumably they had homeworlds, and those homeworlds might have

---

[9] In a 2010 interview for The Discovery Channel, eminent physicist Steven Hawking declared that, "If aliens visit us, the outcome would be much as when Columbus landed in America, which didn't turn out well for the Native Americans."

natural arcanospheres just like Earth, allowing the development of magic. These might be extraterrestrial archmages, come to conquer.

"Or they might be distractions," Vinnie muttered. "Lights and noises. Watch that hand, not this one."

"What are you saying?" Moomin puzzled. "What shall we do?"

"We have to make contact." Mr Testable decided. He clutched his thermos like a talisman. "One of us needs to speak with them."

"It should be you, young man," Dr Shore told Vinnie. "You're asking all the questions. You seem to think you have some answers."

And he shoved Vinnie hard, out of the circle.

The entities glided towards the jobbing occultist, closing to surround him. Vinnie found that he could no longer move his legs to retreat to safety.

"What are you doing?" Moomin screamed at Dr Shore. "Are you mad?"

"A test," the psychologist suggested. "Let's see what happens."

"Get back inside the Faraday circle, kid!" the American called to Vinnie.

"Can't move," the young magus admitted. "Mr Testable, throw me your flask."

The Nuts-and-Bolts man blinked in surprise. "You... want some Oxo?"

"*Now!*" Vinnie roared. The baffled UFO-spotter tossed his thermos to the jobbing occultist.

Vinnie unscrewed the cap. The entities reached for him. "Nutritious beef broth," he told them. "Right in there." He proffered the flask and performed a binding incantation.

It was simple enough, given the raging magics available to shape. The beings were almost made of hunger, and though they wanted Vinnie, Mr Testable had used his old thermos a lot and he believed its contents very tasty. Vinnie amplified that belief and projected it at his adversaries.

Vinnie believed in exorcisms and spirit-bindings; it was on his poorly-printed business card. He had techniques to displace or expel unpleasant and unwanted spirits, including methods of locking them away in objects. Bottles worked well, but in this circumstance a silvered tin vacuum container was best of all.

Three shocked-looking entities disappeared into a foot-long flask. Vinnie put the cap back on.

"What?" Dr Shore frowned. "What did you do?"

"Moomin, come out here," Vinnie instructed. "Right now."

"Out of the safe zone?" the young skywatcher objected.

"Come to me. Please."

Moomin looked at the still-looming red light that had disgorged voracious aliens, and at the scruffy stranger who was beckoning her. She chose to step over the copper coil and join him.

"Thanks. Mr Testable, come and hold your flask, will you?"

The fussy UFO-spotter declined to recover his supper. "You put those... those things in my thermos?"

"Yes. It'll keep them warm. I don't know what kind of Close Encounter this counts as. The Eighth Kind, keeping them in your lunchbox? Anyhow, come and stand here, please."

"Wait!" Dr Shore objected, but by then Mr Testable had gone to join Moomin.

"And now you, Mr American," Vinnie called.

"No," Dr Shore objected. "I think *not*."

The American squealed as if he had been stabbed. He staggered to the ground and clutched his head. "The needles!" he screamed.

Vinnie could move again. He caught the American in a rugby tackle, wrapped arms around his target, and rolled with him over other side of the copper circle.

Dr Shore sprang to intervene. Vinnie got mostly clear, but the psychologist caught his ankle before it was over the barrier.

The jobbing occultist released the trembling American before being dragged back inside the ring.

Vinnie stared up at Dr Shore. Now that he looked properly he could see the same gnawing hunger and utter hatred within the academic as he had sensed before in the glowing entities. "So you were the one who set up this storm of circumstances!"

"And if I was? I'm a psychologist. I experiment."

"No, I'm not talking to Alwyn Shore right now. I mean, he probably did set up some kind of psychological experiment here to test his psychosocial theories, probably fixed on Carol Chanter as the test subject without her ever knowing it. Researchers can be right idiots sometimes. But I'm talking to whatever-it-is that's pulling Shore's strings right now. Because something is."

Vinnie was pinned down by the psychologist's mass, which was much heavier than it should have been. Shore's head twitched from side to side. The motions were not quite human.

"Break the wire," Shore commanded de Soth. "Tear up the tent pegs – or it will go amiss for you."

"Get off him!" Moomin cried at the possessed academic.

"Don't cross the wire!" Mr Testable warned her urgently. "I don't think he can pass over the wire! He's stuck in there."

"He got in by himself!" Moomin objected.

"He was invited in," Vinnie managed to explain despite the pressure of his chest pinning him down. "You urged him to step inside, Moomin. Nobody invite him out again."

"Break this restraining circle or die!" Dr Shore threatened.

"I know what you are!" Vinnie called out triumphantly. "You're what Tom Graves called a *ghoul*."

"A ghoul?" The entity inside Shore snorted scornfully. "I am no ghoul!"

"Of course you're not. I know ghouls. I dated one, and I still have to take her shoe-shopping. I said you're what Graves *called* a ghoul. In the 1970s he wrote a book called *Needles of Stone*, an early look at 'earth energies' using archaeology, geomancy, physics, ghost-hunting, parapsychology...[10] Look into your host body's store of parapsychological knowledge and you'll see what I'm on about. And Graves co-opted the term 'ghoul' to mean a negative manifestation of earth magic, ley energies or what-have-you. All the stuff that *The Ley Hunter* magazine's Dragon Project[11] was investigating. So that's what I think you are. Am I right?"

Vinnie remained pressed under the impossibly-strong academic. He knew his only chance was to keep the indwelling sentience talking.

Moomin shied away from the slouching figure that no longer resembled a noted scholar. "Wait! There's something inside him? Something not human?"

"Remote neuropathic programming," the American insisted. He lay on his side on the downed corn, still recovering from his Close Encounter of the Ghoulish Kind, but he felt he had to explain what was really happening to his ignorant Old World compatriots.

"Not human?" Dr Shore scoffed — or perhaps smacked his lips in anticipation of a meal. "Why would I want to be something so small and insignificant?"

"Well, you seem to prefer wearing that flesh," Vinnie pointed out. "I imagine it's quite different from being a knot of telluric energy that got

---

[10] An updated edition of Graves's work, one of the foundational documents of modern geomancy, is now released as *Needles of Stone Revisited*, Gothic Image (Glastonbury, 1986) ISBN 0-906362-07-5

[11] In 1977, Paul Devereux, the then editor of *The Ley Hunter*, inaugurated the Dragon Project "to look for physical evidence of earth energy/currents which was an axiom of the concept of ley lines." Graves, Devereux, and a succession of researchers and theorists developed much of the thinking behind "modern feng-shui" and many key New Age concepts.

clogged somewhere in an ancient and ailing ley system."

"Alien possession," Mr Testable proposed, pale under the dulling red skylight but determined to make some classification. "You called it a ghoul…?"

The entity in Dr Shore was stung by the description. "I am no pathetic flesh-devourer! I am no star-lost tourist! And the proper name for me is *Grue*!"

"Grue?" the American queried. He was of an age to have played the computer text games *Dunnet* and *Zork*, where the player was often warned, 'It is pitch black. You are likely to be eaten by a grue.'[12]

"From the Scottish term for a shudder or flesh-creep," Vinnie footnoted. "Scandinavian origin; giving us the Old Swedish *grua*, Old Danish *grue*; German *graven*, Dutch *gruwen*, to abhor. In English it gives us 'gruesome'.[13] But really it's a knot of ley energies that has been neglected so long that it's gone waaay off the reservation."

Moomin frowned. "Carol wrote about that. About how the old straight tracks were being lost to housing developments, how the ancient stones had been ploughed away centuries ago, or moved by road construction, or surrounded by power pylons and mobile phone masts."

Vinnie knew that the ley network had co-opted the National Grid and the phone lines decades ago, but he didn't feel it was a good time for a lecture, being as a powerful malevolent force was pressing down uncomfortably hard on his rib cage. Suffice to say that many mobile users were going to find 5G interesting.

"You can call yourself Mr Shiver if you want to," he told the indwelling sentience that had colonised Alwyn Shore. "You can surround us with ley energies so we see lights in the sky, you can divide yourself up to make faux-aliens to torment us and keep us baffled. But now I know that you're a daemon, a creation of telluric magic set as a guardian over a badly-damaged arcane power grid."

Vinnie remembered his father and uncle arguing over the dinner table about whether the European ley system required systemic repair or planned

[12] Jack Vance revived the archaic word 'grue' in *Eyes of the Overworld* (1966), and it has since been restored to general usage. Dunnet, abbreviated from 'Dungeon Arpanet', was an early text-based computer adventure which featured grues as monsters; it is from this game that the American's quote came. The grue was ranked 46th in IGN's 2010 list of Top 100 Video Game Villains of All Time.

[13] In some parts of the UK dialect speakers will still say of something that ran a shiver up their spine, "It gives me the grues."

destruction now that it was so degraded; just before Asteroth de Soth had made swarms of maggots gush from all of Uncle Belial's orifices.

"If you comprehend my nature then you know that I will gladly torture you until you release me," the Grue snarled. He leaned harder on the jobbing occultist.

"I also know why you're getting weaker. Sure, you're scary-strong holding me down now, but inside this ring, cut off from all that raging power that was sustaining you, swelling you up... I don't like your long-term odds."

"I'll kill you."

"You could. Folks, listen. *Nobody* lets him out of the circle. Not for any reason. He can't last long in there. Eventually Dr Shore will take control back."

"I'll kill Shore too," Dr Shore's mouth promised.

"Still doesn't get you loose," Vinnie countered. "You didn't plan for my circle and you didn't plan on getting trapped in there. You don't need a lot of imagination to borrow human images of UFOs and aliens, so we'll overlook the lack of longer-term planning. Bottom line is, you can kill me, but then you starve to nothing. Or..."

"Or what?" the Grue couldn't resist asking. It was newly sentient, still relying on its academic host's cognitive capabilities. It hadn't yet learned not to listen to plausible magi.

"Or we make a deal. You cough up Carol from wherever you transfigured her to feed off her. You call off the *X-Files* episodes. You return into the telluric tides and stop bothering people. Basically you just *back off*."

"*Or* I tear you to pieces and devour you body, mind, and soul, and grow strong enough to burn away this little ring of annoying metal," the Grue counter-offered.

Vinnie snorted scornfully. "You keep trying to make us look in the wrong direction, don't you? You're quite deceptive for something that must have originally been designed as a system guard dog. Hardly magic at all, just nature channelled. But think about this: it's about six thousand years since the geomancers installed your ley system. That's a long time to go without a patch. And that's millennia for mages to evolve magic-working from cutting people open in henges to... well, I won't say more civilised occult practices, but at least more sophisticated ones. You, Mr Shiver, are very outdated."

"And you will soon be dead," Shore's lips promised.

"You're claiming he's not extraterrestrial at all?" Mr Testable asked, disappointedly. "Extradimensional, then. Ultradimensional?"

"You're big and powerful and fierce, and rather mean and nasty," Vinnie told the Grue, "but we've learned a few tricks since your time. I was able to bottle your spawned-off soldiers, wasn't I? Well, thermos them. That had to be a bit of a shock. Same with extruded copper wire and camping accessories."

The jobbing occultist felt just the tiniest bit sorry for the manifestation. It had awoken after so long, a last synapse flaring in a broken network, and it didn't understand anything at all. It only knew an overwhelming malice towards its creators who had moulded it into existence and then abandoned it, and an endless hunger to be filled.

"Why does it hate us so?" Moomin pleaded to know. "I can feel its hate."

Vinnie could guess. "It used to get fed regularly, to keep it strong and vigilant. Neolithic blood rites, I think. The Druids weren't squeamish about rehabilitation of offenders. But then the care packages stopped. The instructions ceased. Everyone just forgot about the Grues, abandoned them. Left them to starve, to be useless."

"You're very clever for a mortal wisp," Dr Shore's driver admitted. "Cleverer than my shell, who thought himself to be devising a noteworthy social experiment in shaped collective perceptions."

"Which you borrowed and used for your own devices to wake up and feed. I need you to cough up Carol now, Grue."

"You're clever, as I said. But clever isn't enough." Dr Shore leaned in with ribcage-cracking force. "Now *this* is a close encounter! Break the prison."

Vinnie screwed off the top of the vacuum flask. The energies of the three captured 'aliens' seared back into the Grue, stunning it for a moment.

"Fool!" the Grue thundered. "You have only made me more powerful!"

"'Cause I have," Vinnie assured the entity. "You can't carry out commands if you don't have the telluric energies for it."

"Commands? None command me."

"Well, none have commanded you for a really long time, but that doesn't mean the old pathways aren't still there." Vinnie fumbled into his pocket and held out his hand. "Here."

The Grue actually shied back a little. "What is it?"

"Food. Moomin's mint, actually, but that doesn't matter. It doesn't have to be blood sacrifice really. Just an offering. A fee. And you are *very* hungry."

The Grue didn't want to accept the food. It didn't have a choice.

"It might not be the wages you want, or even need, but it's what you've been offered," Vinnie told it. "Now hear your instructions."

"You can program it?" the American wheezed.

"It can reveal ancient wisdoms?" Mr Testable hoped.

"It doesn't have any ancient wisdoms," Vinnie cautioned. "Only implanted imperatives put there so long ago it doesn't even remember them. Even thinking is new to it right now. So Grue, here's your work: return Carol Chanter to us, as unharmed as she can be. Gather up whatever energies you've sent sloshing about and shut down the UFO cosplay. Quit Dr Shore with as little damage to him as possible. Then dismantle yourself and go back to sleep."

"Or what?" the Grue challenged. Spittle foamed from Dr Shore's mouth.

"Or I'll pull out this tent-peg," Vinnie threatened.

"You said not to do that," Moomin reminded the young magus worriedly.

"Change of policy. Grue, will you obey my lawful order? You took the mint of offering. Will you now deny the service you are obligated to render?"

The guardian of the leys was too soured now to co-operate. "You are not one of my creators. Even if you were I would slaughter you for what you have done to me. I will not rest, will not dissolve. It will be my joy to dismantle *you* instead!"

"Then it's time for the peg."

Vinnie snatched up one of the metal spikes that insulated the circle about the Grue from the tempest of Beltane energies it had provoked and harvested. The entity was once again suffused with the vortex around the crop circle, with the pulsing energies of the Ridgeway track, with the telluric streams from nearby Silbury Hill and West Kennet Avenue, the Avebury stones and Green Street. Power surged and sang over the whole of Wiltshire's Neolithic landscape, from Stonehenge and Woodhenge, from Glastonbury Tor and far beyond.

The dull red sky-light flared bright as day. Similar lights danced around menhirs and stone rings from the Orkney Islands to Southern Brittany. Across the whole of Europe the straight tracks and fairy roads and corpseways glowed.

A harsh wind blew up, almost bowling over the watchers still in the cornfield, pressing down the cereal stalks in a new, much vaster spiral.

"I warned you," Vinnie de Soth told the Grue.

"Warned him what?" Mr Testable demanded, struggling to stay upright.

"Before, this Grue could draw on all this power to do whatever it wanted. But now it has been given a task and it has denied it. It has malfunctioned." Vinnie gritted his teeth against the tempest. "And it is not the only Grue."

Moomin knelt over the American to shelter him from the rising chaos. "You woke the others!" she gasped at Vinnie.

"He did what?" objected Mr Testable. "How? Never mind that… *why*?"

Vinnie's shaggy hair whipped in the wind. "Well, I arranged to get their attention on the situation here. Let's hope they've not all been twisted beyond recognition, eh?"

The vortex-thrashed corn bent over in complicated patterns. The lurid red illumination over the site was supplemented by other luminous aerial glows, a profusion of coloured lights and rippling images. Spokes radiated out from the hub, shining shimmering dead-straight lines vanishing off into the distance to the furthest edges of the British Isles and beyond.

From Clava Cairns to St Michael's Mount the leys awoke on Beltane, and with them woke the guardians.

One grue was different from the rest. One grue was not just terrible but *wrong*.

"You might want to grab a tent-peg fast," Vinnie called out to the UFO-spotters. "You really want to be grounded right now. Earthed."

Because just then, everybody there was either with the earth energies or against them. And the planet was going to win.

Shore trembled as if he was having a fit. Vinnie seized up the empty flask and crammed the academic's business card through the thermos' neck. "Possess this aspect of him!" the jobbing occultist demanded.

The fury of the other guardians demanded obedience.

The twisted Grue wailed in hatred and despair as it was ripped from the body it had taken and was compelled into the magician's vessel.

Vinnie tried to make a quick calculation of the magnitudes of forces swirling around now but only got a headache. *Lots*, was his general conclusion.

"Sorry about your Oxo," he called to Mr Testable. "I need it for the exchange."

"What the hell are you talking about?" the American demanded, sheltering under Moomin from the uncanny storm and the pulsing lights.

Moomin glimpsed Vinnie's intent. "You mean to exchange the... the Grue for CallMeOsgood7, for Carol. These... these ley things could swap her for the Shiver?"

"That's the plan. There's always got to be a balance. Destroy the Grue, return the mortal, disperse the phenomena, with enough spare energy for a big explosion left over." Vinnie dived at the confused, wakening Dr Shore, tackled him to cover, and grabbed a tent-peg. "Everyone duck now!"

There was an actinic flash and a radio pulse that set off military alerts across Northern Europe and would launch a thousand conspiracy theories and UAP sightings.

Vinnie's only part was to tweak the exchange of energies. That was what a magus did, of course, but this was more like wiring a power station than a three-pin plug. What it mostly meant was "don't get it wrong".

Darkness and quiet descended on Goreston Farm.

Vinnie helped Dr Shore up to his knees. "What... what happened?" the psychologist asked blearily.

"Your unethical psychosocial manipulation trial got out of hand," the American accused him.

"I shall sue," Mr Testable warned him. "And you owe me a thermos flask."

"Uh-oh," Vinnie realised. "Dr Shore, I used your business card for affinity to drag the Grue out of you, into the bottle. It was destroyed along with the Grue by the, er, the other Grues. It had your credentials and job title on it. It's just possible that it might have taken your academic career with it. A sacrifice."

Moomin rose blearily and made another discovery. There was another person with her in the greatly enlarged corn circle. "Hey, are you Carol?"

A confused, thin-looking woman in a knitted bobble-cap looked around with wide, horrified eyes. "What happened?" she whispered.

"Nothing you'll remember," Vinnie promised her. He hadn't traded for her memories of the Grue to come back with her. "Your mum is worried about you. Call her."

"It's a good question, though," Mr Testable echoed the rescued skywatcher. "What did happen here, young man? And how do you know so much about it?"

"Oh, I'm not allowed to say," Vinnie assured him. "The Men in Black, you know. It'll all be covered up."

"I knew it!" the American celebrated.

Any actual cover-up was likely to come courtesy of the Nine Great Houses of the Western Occult Tradition, and especially House De Soth who were rather embarrassed by their wayward son, but Vinnie was content to leave the saucer-seekers with their theories. Everyone should have theories.

"They follow the old straight tracks," Carol confided in Moomin. "I'm sure of it now."

Vinnie slipped away as they began to argue that statement. He warmed his hands on the remains of the Beltane fire and then hiked off before anyone turned up to complain about the fireworks display.

And if that pattern of ley energies really did attract beings from another planet, Vinnie was content that someone else could take a shift to deal with them.

TAHDUKEH

CARSTEN SCHMITT

The boy's body smelled of the river from which he had been recovered a few hours earlier. The odours of water, faeces and the effluents from the ironworks along the Sâr blended with the vapours seeping out from the mouldy walls of the cellar vault into a peculiar scent.

"Well, who's that they assigned us, Pit?"

Pit's nose was red from burst veins under the skin. The old gendarme scratched it with the dirty nail of his index finger.

"Someone fished him from the Sâr this afternoon at the coal weighing station, Commissioner Charrois. At first, they thought he was a barge-tower's lad who fell into the water, but then they noticed the wound on his head. We don't know who he is." he concluded succinctly.

"And that's when someone thought of us? What do you think, Pit? Is it going to be a big case?" Charrois laughed dryly, but Pit kept his stoic seriousness: "Can't tell. He had nothing on him. Maybe he came from a public house and got mugged."

Through a small window high in the basement vault of Sârbruck's gendarmerie station shone a few last rays of sunlight. Charrois threw back the grey wool blanket that covered the boy's body. "Let's get to it, then. Light the lamps, Pit, be so kind."

Pit lit the oil lamps hanging on chains from the ceiling, and the body was bathed in warm yellow light. *"A beautiful boy,"* Charrois thought, *"almost a man, shaped by physical labour but not yet broken from it"*. He looked at the face framed by blond curls. In a few years, the boy would have cut a dashing figure in the blue and red of the ducal guard. Charrois pulled the boy's jaws apart and bent over the open mouth. He pressed on the chest and belly of the dead body and smelled the air gurgling from his lungs. "If he was in the inn, he didn't drink much."

Then, he took the boy's head in both hands and turned it. Charrois touched the swollen red wound which lay hidden under the blond hair at the back of the head.

"This looks like the mark of a crude weapon, maybe a cudgel or a rock."

Pit shrugged his shoulders. He kneaded his hands and stepped from one foot to the other. Charrois pulled his watch from his vest pocket and looked at the dial; half-past eight. Pit was getting restless.

"Will the Commissioner still need me?"

"No, Pit, it's fine. Just put my chest on the table there and be back in time to pick me up tomorrow."

"Very well, Commissioner." The gendarme saluted and left the vault, closing the door behind him.

Charrois trusted Pit. The man would usually be still half-drunk in the morning, but they had an agreement. Neither asked questions about the other's weakness, and both made sure that no one knew more about them than necessary. He turned to the small chest that Pit had placed on a table against the wall of the vault. Charrois opened the lid and felt a tingling in his hands at the sight of the props that were neatly lined up inside.

The commissioner lifted the tray of black-lacquered wood from the case and placed it on the stretcher next to the lad's body. He methodically inspected the equipment. The oil lamp with the bonnet made of crystal glass, the bamboo pipe with the white porcelain head, the deep spoon of hammered copper and all the other small tools, bottles and containers his vice required.

Charrois lit the lamp. To him its yellow light was like the flame of anticipation that would light his way into the dreams of the past. He took the vial of smoking opium and drizzled a few drops of dark brown chandu into the spoon.

Then, he took a tiny jar of ebony and opened it. When he saw the paltry amount of fine grey powder inside, he hesitated. Was he willing to sacrifice his dwindling supply for a slain peasant bumpkin? His Tahdukeh, his wonder drug? But to stop now would have been just as impossible for him as pausing the beating of his heart.

He measured a small amount of the powder with a spatula and mixed it with the thick brown drops at the bottom of the copper spoon. Then he held the spoon over the opium lamp and boiled the chandu until it became a sticky mass, which he rolled into a plug that fitted exactly into the opening in the head of the pipe. He lay down on the stretcher, the lamp between him and the corpse, and with bent legs turned onto his left side and rested his head on his rolled-up coat. With his right hand he held the pipe over the flame with the opening facing down and started inhaling the delicious vapours.

Charrois slid into a state of serene attentiveness, free from the drab dullness of life. The boy's body lay opposite him, his head tilted towards him and his eyes opened to slits. *"You will not be buried a nameless dead; we shall find out who you are,*" Charrois thought. He sat up and studied every square inch of the naked corpse until he stopped at the boy's right hand. Charrois reached out with his index and middle finger to stroke the corpse's palm. His fingertips felt hard calluses and his own hands tickled in response. The first wave of Tahdukeh rapture was approaching and Charrois awaited it with his eyes closed. He let it wash over him and whirl him through an invisible maelstrom until the dizziness subsided and he felt solid ground under his feet again.

The vision was not stable enough to open his eyes, and so he had only the remaining four senses left. The stink of the cellar had given way to the animal smells of a stable and horses. In his hands he felt a wooden shaft like that of a broom. No, not a broom, the object was too heavy for that. It was a shovel, and with this realisation the sensations that had besieged his senses exploded in his head. The smell of horse and stable, which he had perceived earlier, became overwhelming. He heard the scratching of the shovel blade on the ground and the damp hissing sound when it stabbed into the dung while horses snorted around him commenting on his work.

"Mattel!" He jerked at the sound of the voice. He opened his eyes and looked around for the caller, hurling from the vision. The dream image had not been consolidated enough, sight been too much for it. He had glimpsed a stable with whitewashed walls. The horses, all well-fed animals, were housed in neat boxes. This wasn't a coaching station, or a coachman's stable, but that of a noble gentleman.

Somebody had called a name, Mattel. Was it the dead boy?

Charrois prepared another pipe. This time, he balanced the mixture of Chandu opium and Tahdukeh perfectly, so he awoke in equal parts in the real world, and the vision of the drug. He saw the young man's body lying naked on the stretcher in front of him, while he was looking through his eyes at the same time.

Mattel had leaned the shovel against the wall and had gone to the stable door from where the stranger had called. Like a drinker who lets the wine roll over his tongue to taste the bouquet, Charrois traced the sensations of the man whose soul he inhabited. He sensed joy and excitement, but also familiarity. The caller had been no stranger to Mattel.

He stepped out of the shadow of the stable into the yard and squinted his eyes in the bright sunlight. He saw the surrounding buildings only as blurred white and red surfaces of their outer walls and tiled roofs.

Mattel's attention was focused on a man strolling towards the edge of the forest. He wore a white frock coat decorated with iridescent green and blue Cathay-style embroidery, as well as trousers and stockings of white silk. His feet were in shoes with golden buckles and in his hand he held a delicate walking cane. The stranger's face remained hidden, but Charrois sensed that Mattel knew who he was. The man slipped into the forest and with his disappearance between the trees Charrois felt the vision slipping away from him. He was dizzy and instead of the bright sun in the sky he saw four points of light from the oil lamps hanging from the ceiling of the basement vault.

With the practiced fingers of a Tahdukeh dreamer he groped for his smoking implements, prepared the pipe and took a drag. The dizziness subsided and the image before his eyes stabilised but remained motionless. Following his intuition, he dared to send his spirit back to the cellar of the gendarmerie guard house. He took another puff of the pipe and as the drug whispered to him he followed that intuition further, bending over and kissing Mattel's dead lips.

* * *

When Mattel parted his mouth from the lips of his counterpart and opened his eyes, Charrois looked through those eyes into the pretty face of a beardless young man.

"Mattel, my dear Mattel." The other wore a wig whose grey powdered hair contrasted with his dark brows and fawn eyes. Charrois, a witness of what Mattel saw, thought he knew the stranger. He grasped for the sensory impressions and fragments of memory he picked up, but the name escaped him until Mattel spoke.

"My prince, I…"

"Don't call me that, Mattel. I'm your friend."

Charrois could feel Mattel smile. It was not the first time they've had this conversation.

"I've got to go. If the stable master notices I'm gone, he'll cuff my ears."

"He shall not dare!"

"Henrik, out here you're my friend. But back there, you're the prince, and if you concern yourself with a tardy stable boy, people will talk more than they already do."

"Let them! They won't utter their ugly thoughts to my face."

"But to mine. And worse."

"Don't worry, Mattel. I won't let anything happen to you."

"Nevertheless, I shall go. It's time you were a prince again and I a groom."

Henrik of Sârbruck, hereditary prince of Duke Ludovic, nodded. They rose and Henrik picked up his coat, which he had spread out as a blanket under them.

The pleasant faintness of the opium enveloped Charrois, and he threatened to slip into sleep, but until he had revealed the last secret, he could not rest. His hand wandered back over Mattel's cold cheek to the wound at the back of the head. His index finger pierced through skin and crushed bones into the cavity of the skull. He opened his mind for one last vision.

* * *

The man with his dark brown coat and hat of the same colour could hardly been seen among the trees. Mattel recognised Johann, the Court Marshal's lackey, and this realisation sparked discomfort, even fear of the piercing dark eyes and the smile that distorted the lips of the hard mouth and revealed a set of broad teeth.

"You're Mattel, are you not?" Not waiting for an answer, the man grasped his upper arm with a firm hand. The grip itself was not uncomfortable, but it held a veiled threat that this might change at any time.

"His Highness has instructed me to bring you to him."

"The prince? We, I mean... why does His Highness wish to see me?"

"His Highness has taken me into his confidence. You need not fear."

Why did the man think he was afraid? Mattel looked into Johann's eyes and saw he was lying. The heavy hand on his upper arm increased its pressure. But even if he had run away, where would he have run to? Johann knew him, knew where his mother and little sister lived. Charrois sensed resignation rising in Mattel. Johann pointed with his chin: "That way, and slowly. We wouldn't want you to fall and hurt yourself."

Mattel turned around and walked in front of the man. Blood was throbbing in his ears and his sweat-soaked shirt stuck to his back. *'Don't let him slaughter you like an animal!'* Charrois thought and as if Mattel had heard his thoughts across the border to the next world, he bolted. His strong, young legs were used to covering long distances, and through the bare soles of his feet, which had never known boots, he was connected to the earth, sensing roots and treacherous snares in advance, jumping over them, zigzagging like a wild hare. The more of a lead he gained over Johann, the more his fear subsided, and he found his pace. Quickly, he made his way through the thicket, never running faster than his endurance allowed. Soon he could no longer hear the heavy-booted footsteps of his pursuer, or his cursing when a thorn caught his coat.

How foolish it had been to think of surrender! Johann might know who he was and where his mother and sister lived but never would his master, his Henrik, allow any harm to come to him or his family. If he secretly returned to the castle and sent Henrik a message, he'd be safe. A grin crossed Mattel's face as he thought about how he would get one over on Johann the Bloodhound.

He was still laughing when he lost his footing and his momentum carried him over a steep slope that the undergrowth had hidden from his view. He rowed vainly with his arms and legs and screamed as he hit the ground. His right knee burned as if a carpenter's nail had been driven through it. When he tried to stand up, his leg failed him. Mattel collapsed and buried his face in the leaves. No scream, no crying. Only luck and secrecy could save him now.

He moaned when he heard the dull sound of a heavy body landing on the ground. Steps approached the spot where he lay and soon a hard boot heel on the back of his head pressed his face deeper into the dirt.

"What a swift little doe. Or is it a buck? Tell me, did he offer you his arse or was it the other way around?"

Mattel's mouth was full of earth and rotten leaves that suffocated any response. He struggled to get up, but a sharp kick of Johann's boot sent him back to the ground, bringing tears to his eyes. "I wanted to cut your throat, but I have a better idea. Instead of sullying my good knife with your sodomite blood, I'd rather bash your skull in." The pressure on his head eased as the man lifted his foot. Mattel turned his face and spat leaves, but even that movement caused him such pain that he couldn't think of getting up. He saw Johann bend down and pick up a stone bigger than his fist. The air was forced out of Mattel's lungs when the heavier man returned and knelt on his back. Pain, sharp and clear like a flash of ice, exploded in the back of his head and seared through Mattel's mind, letting him perceive the world for a single moment in unnatural clarity. He saw every detail of the forest floor before his eyes, every leaf and every single seedling that crawled like white maggots from the acorns and beechnuts of last autumn. He smelled the cycle of growth and decay, the carcass of an animal rotting in the undergrowth, as well as the fresh green leaves sprouting from their buds.

In his ears, all sounds mingled into a tormenting cacophony: Johann's panting breath, his hot drool dripping onto Mattel's neck, every crackling and rustling in the bushes, the singing of the birds and even the Sâr, which ran its course in the distance at the foot of the Halberg. Her gentle murmur filled his mind, becoming one with the thunderous pain that surged through his head before it broke on the cliffs of his brow.

* * *

Charrois opened his eyes and screamed. He tasted phlegm and salty tears on his lips. His head was ringing like a bell and his limbs were icy cold.

The door to the vault opened, and he heard footsteps on the stone floor. Then, a calloused hand covered his mouth, and he smelled Pit's liquor-filled breath and heard his hissing voice: "Hush, Commissioner! Everyone can hear you throughout the whole cellar. The Examining Magistrate is coming with the boy's murderer!"

Charrois stretched his limbs and let Pit help him up. He rubbed his crusted eyes and with the gendarme's aid stowed away the smoking utensils. He straightened his coat and brushed a curl from his forehead when Examining Magistrate Carl Theobald entered the basement vault. The judge bent down so as not to hit his head on the lintel, but once inside, he straightened up to his full height again and took in the scene with a sweeping glance. Pit, standing there with his hands crossed, tried to block the view of the small chest. Charrois, sweaty and with dark-rimmed eyes sunken in his pale face, tried to look as if he had just started his shift.

"Good morning, Coroner."

"Commissioner Charrois," Judge Theobald didn't bother with pleasantries, "the matter here is settled. We have the murderer." Behind Theobald, his secretary had entered the room, along with two gendarmes who dragged a battered man between them. One of his eyes was swollen and a trickle of coagulating blood ran from his shattered nose into his matted beard.

"Jakob Omlor, does he see before him the body of Matthias Marx, known as Mattel, whom he has slain?" The man stared at his feet but nodded. "Make a note: 'The body was presented to Jakob Omlor, who again confessed in the presence of witnesses.' Get him out!" Theobald nodded at the gendarmes, who led the unfortunate man out of the vault.

Charrois was stunned. "Begging your pardon, Coroner, but who is that man?"

"A drunkard and a thug. Marx caroused and diced with him the night he was killed. Won too much, so Omlor suspected foul play and wanted his money back. He killed him in anger and dumped the body in the river. The canaille confessed immediately when he was apprehended."

The expression on Theobald's face forbade further questions. "Go home, man, and get cleaned up. The way you look, you're a disgrace to the gendarmerie!" Theobald turned to leave when Charrois noticed that one of the Coroner's leather gloves, which he had stuffed in his coat pocket, had fallen out. He bent down and picked up the glove and gave it to Theobald, who accepted it with a wordless nod and left.

Pit covered Mattel's body and asked, "Now what, Commissioner?"

Charrois touched his right hand. For a moment, when he had picked up Magistrate Theobald's glove, he had sensed how the hand with the glove had received a purse; a purse heavy with coins with the ducal crest on it.

"You heard the Examining Magistrate. We're going home."

THE DEAD SHALL RISE

C.L. RAVEN

I bring the dead back to life.

Then I watch them die again.

Some people think I'm in league with the devil. Some think I *am* the devil. Nowhere is the devil depicted as a petite, tattooed, pierced goth woman with black and purple hair. Trust me, I've studied Demonology. People get upset when I tell them the devil isn't real and the cruelty of the world is down to humans. They want someone to blame. But the devil isn't murdering people.

That's where I come in.

I'm Indiana Raine. Necromancer Detective.

* * *

Creak.

Creak.

Creak.

Rain punished the earth, the sky mourning another life taken. Ancient trees whispered in a language our ancestors forgot. Police tape formed a barrier between the living and the dead, the crime scene resembling modern art roped off from admirers. Death didn't come with a price tag. Night's midnight shroud cloaked everything beneath it.

Including the gibbet cage hanging from the oak tree.

A floodlight clicked on, exposing the dome-shaped cage. It reminded me of a birdcage. Except the occupant wasn't perched on a wooden swing singing to be set free.

A skeleton clutched the bars, its mouth open as though it died screaming. It probably did.

Forensics gathered evidence into pots and bags. One used bolt cutters to remove the padlock sealing the cage.

I stopped beside a tall, toned man wearing black jeans and a distressed look jacket. He had light brown skin, close cropped hair and startling emerald eyes.

"When you said 'strange', I presumed occult symbols painted in blood. You undersold this."

"Anti-climaxes are disappointing. What's your opinion?"

Jaxon Graves was the detective I liaised with. The others found me creepy. He was decent enough to try to get past that.

"He died elsewhere."

"What makes you say that?" His soft Welsh lilt betrayed his west Walian ancestry.

"Miles of woodland and he left it in the public area. I know society is glued to their phones, but a skeleton in a gibbet cage is hard to miss. If he'd died here, someone would've Instagrammed it."

"I'll scan Instagram for hashtag skelfie."

I cackled.

"A dog walker found him."

"I should get a dog. Apparently I need a hobby. Corpse Finder General sounds like an interesting one. Something to liven up my journal."

A PC stepped backwards. Some people disliked ambition in others.

The cage rotated, the branch creaking. The floodlight caused shadows to dance in the skeleton's eye sockets, as though its soul was trying to escape.

"Were they alive when they were put in the cage?" Jaxon asked.

"Why build a torture device for a corpse?"

"*Torture* device?"

It may have been the moonlight spying on us, but he matched 'Goosebumps Grey' on the Freaked Out Colour Chart.

"Someone didn't pay attention in History."

"And you were the model student?" he teased.

"I got an A in medieval punishments."

He paled three shades to 'Ghostly Pallor'. One day I'll scare him to 'Alabaster Heroine'.

Forensics lowered the cage. Jaxon handed me a white Tyvek suit, gloves and bootees.

"Do your magic, Lazarus."

"You might want to leave."

"I've seen you wake the dead."

"You saw me wake a freshy. You can't un-see what I'll show you."

"I'm staying."

"Fine, but I'm not holding your hair when you're puking your supper."

I pulled the suit over my black satin skirt and corset, slipping my mesh covered arms into the suit's arms. Perks of not being a police officer – I can look stylish while solving crimes. Like they do on TV. I raised the hood and donned the gloves. My boots bulged out under the legs and bootees. I

191

approached the cage, pulling black candles, a bottle of ink and a flax leaf from my bag. I arranged the candles in a pentagram beneath the cage and lit them, focusing on the flames. Waking the dead was a complex art. If my mind wasn't clear, I could bring anything back.

Or anyone.

I gripped an almond-wood trident I'd carved from a tree with a virgin knife. Witchcraft wasn't cool kinetic tricks and fire powers. It had rules. Consequences. The ink was a blend of red ochre, burnt myrrh, fresh wormwood juice and powdered evergreen leaves. Dipping my glass pen into the ink, I wrote AZEL BALEMACHO on the leaf. Forensics eyed me warily. This wasn't their kind of science.

The door squeaked as I opened it. I placed the leaf in the skeleton's mouth and rested my hand on the skull.

"*Mortuis resurgemus.*"

The dead shall rise.

My fingertips tingled. Darkness overwhelmed me, drowning me. Discomfort and hunger haunted me. Pain hammered on my skull. My heart thumped my ribs, trying to escape its own cage. Muscles, tendons and ligaments snaked around the skeleton. Blood surged through empty veins, turning them blue. Flesh crawled over the bones, the skin forming last, sealing the horror within. Blue eyes stared from a terrified face.

He woke screaming.

Jaxon photographed the victim, fighting to hold the phone steady.

"I'm Indiana. I've temporarily raised you from death, but you cannot stay. What happened to you?"

Images bulleted my brain. A cloaked figure. Gibbet cages hanging in a dim room. Intense hunger pangs. Cries for help. Despair. Fear. Pain.

I pulled my hand away. An anguished cry escaped the man then the body decayed, returning to the skeleton.

I extinguished the candles and placed them into my bag, my hands trembling. Jaxon handed me a tissue. I wiped blood from my nose. Necromancy took a toll. I had to give something of myself to get something from them. Sometimes, they took too much.

"He was alive when he was locked in there. He starved to death. There're others."

"Did you see the killer?"

"Someone in a hooded robe. If this is another Cthulhu-raising cult, you're on your own. Being a sacrificial offering is something I've ticked off my bucket list."

I shed the suit and stumbled to my hearse. I switched the stereo on to blast Disturbed then climbed into the back and closed the curtains. A necromancer having a hearse is cliché, but they're practical and designed for people to lie down in, which saves money on hotels. Most of all, it provides privacy.

I didn't want anybody witnessing the aftermath.

I curled into a ball, racked with pain. Tears spilled onto the wood beneath me, yesterday's mascara bleeding prison bars down my face. I dug my fingernails into my scalp and bit my lip to stop screams and sobs escaping.

Nietzsche said when you gaze long into the abyss, the abyss will gaze back into you. I touched death. When he touches you back, you die.

Not physically, but part of your humanity dies. Music makes me feel alive, so I use it to exhume my mind from its grave. My best friend and fellow outcast, Saffron, uses sex to drag her from the abyss. My arch nemesis, Dagan the Undead (real name Clive), drinks blood from his vampire groupies. Others seek thrills, food, or fights. Anything to shake Death's hand off your shoulder.

Today, music didn't bring me back. Today, I needed pain.

I sliced my forearm with my dragon claw ring. As blood escaped the wound, it took the darkness with it.

Someone knocked on my window.

"Are you done meditating?" Jaxon asked.

I wiped my eyes, smearing evidence of my pain across my face. "My chakras are aligned and ready for action."

"I'm heading back to run the photo through facial recognition. I'll let you know if I identify him."

A necromancer hanging around the police station tainted their image. They didn't want to be the subject of mocking headlines. I stayed until they brought the skeleton up in a body bag and put it in a black coroner's van. The cage followed, wrapped in plastic. I followed the coroner down the narrow windy country lane. From the mountain, I could see Cardiff, the Bay merging with the sky. The city was waking up.

I drove to my shop, Occulture Shock on the corner of Death Junction. It looked how you would expect an occult shop to look. Dim lighting, black shop front, Ouija boards, crystals, pendulums and incense sticks on display, along with tarot cards, essential oils, spell ingredients, and coffin bookshelves stacked with books on witchcraft, spells, spirituality, astrology, demonology. Shelving units held crystals and rune stones with information cards, and fairy and dragon ornaments. I'd stopped telling people the occult was the study of hidden things, including the paranormal. People believed it was all witchcraft and Satanism. Things they can pretend to love to shock the PTA. Necromancy didn't pay well and I had bills, so I sold them their beliefs. It's how I met Jaxon. A Ouija board from my shop had been found at his crime scene. I'd enjoyed his discomfort.

The skull windchime above the door tinkled as I entered. Saffron leaned on the counter reading a book on astral projection. She was tall with long red hair, snake bite lip rings, earrings lining her ears, a tongue piercing and a septum piercing. Most of her skin was decorated in tattoos. She wore a tight black PVC dress. She had a striking beauty that made straight women and gay guys question their sexuality.

"Been 'liaising' with Detective Hunky?" she teased.

"Been bringing the dead to life."

"He's almost the spitting image of Jesse Williams."

"Skeletons don't resemble anybody."

She shook her head, smiling. "I never get to tease you about your crushes – you never have any. Don't deny me my fun."

I joined her behind the counter. Wands, athames and ceremonial daggers were locked inside it. A display of gothic rings perched at the front. I drummed my fingers on the counter, my rings chinking. I wore one on every finger and had as many piercings as Saffron: lip, nose, four in each lobe, cartilages, tragus and daith.

"Someone starved a man to death in a gibbet cage."

Saffron closed her book. "Bit archaic. And cruel."

"There are more victims."

"Is it a cult? Or a weirdo with a fascination for medieval torture? No wait, that's you." She laughed.

I smiled. "I can't help having interesting hobbies. I saw one person in the skeleton's memory but trauma blocks things."

"Ritual sacrifice? Though they favour bloody offerings."

"Plus with supermarkets, people don't need to worry about a good harvest. Sacrifices went out of fashion when dress sense improved."

"Some cultures and religions still practise it. Though usually chickens are the victims. Cruelty never goes out of fashion."

I fetched my tablet. Religions favoured fasting to get closer to their gods, so maybe some used starvation as a form of sacrifice. I didn't notice Jaxon enter until Saffron made a siren noise with 'incoming hot guy alert'. I laid the tablet down as he reached the counter.

"Most people scroll through social media on their breaks. You're researching torture methods."

"You never know when you need an Inquisitor's Chair. I like to be prepared."

"I know what to get you for your birthday."

"I'd like one in black, please." I tapped the tablet. "Gibbeting was used to display the body as a deterrent. Sometimes people were gibbeted as a form of execution. Last case in Britain was 1834. Bodies might be coated in tar to preserve them so they could leave them up for longer. One man's body hung for twenty years. I bet he became a town landmark. I wonder if they decorated him at Christmas."

He spun the tablet to face him and scanned the medieval drawings. "Why can't I get domestic murders like everyone else?"

"You'd get bored. Plus you wouldn't need me. And I know you'd miss visiting the shop."

He eyed the skull sweet bowl. After hesitating, he took a lolly, unwrapped it and put it in his mouth. "This body wasn't preserved in tar or publicly displayed."

"It was displayed at a popular site."

"In woodlands only locals know about, not hung from Cardiff Castle like the Christmas lights. Are you sure he starved to death? Not put in there afterwards for shock value?"

"Forensics will find traces of bodily waste on the base."

Jaxon took the lolly out of his mouth and wrapped it up. I smirked.

"You didn't get the victim's name when you… resurrected him?"

"No. I can only access their senses and emotions at their time of death. Their memories are disjointed images from a dying brain. You know eyewitnesses aren't reliable. When the body dies, so does the mind. Think of it as rescuing data from a ruined hard drive. A lot gets lost and what's left has been corrupted."

"I remember the last skeleton I resurrected," Saffron said. "I needed a threesome to scrub that stain off my soul."

Jaxon raised an eyebrow.

"The worse the deaths, the harder it is to retreat from the darkness. You see the bodies. We see the deaths. Sometimes, we die their deaths. So we need to recover. We need to feel alive."

"I'll find out who's making gibbet cages."

"Try bondage sites. They're now 'a unique bondage experience'. I've seen them online for a grand."

"A torture device is now fetish equipment?"

Saffron smiled. "Pain can be pleasurable. And who doesn't want a hotty imprisoned in a cage like a sexy bird of paradise?"

Jaxon shuddered. "Lazarus, I hope you're wrong about there being others."

"This is only the beginning."

* * *

Darkness. A lock clicking shut. Cold metal pressed against my skin, sharp edges gouging my flesh and tasting my blood. Weeping. A light flickered on. Shadows twisted up the stone walls, their gnarled fingers reaching to tear my soul from my rotting body. Bars surrounded me. Pungent stenches poisoned the air. Faeces. Urine. Stale sweat. Open wounds. Infection.

Cages hung from beams. Some were empty, their open doors taunting those imprisoned. The floor beneath them was stained. People in varying stages of decay occupied other cages. No, not decay. Dying. Only one was dead. Me.

Wearing someone else's skin is uncomfortable. Wearing it while it's wasted away is worse. Breath struggled to reach my lungs. My stomach felt hollow, my throat dry. Pain pounded in my head; hunger, fear and misery combining to punish me. Strength abandoned my body. My dry, swollen tongue scraped my split lips to wet them but only offered more pain.

"Where are we?" I asked.

The woman near me screamed. "I watched you die."

That can't have been easy. It's rarely as peaceful as films pretend.

"I'm Indiana. I've borrowed the body to find out who's doing this. And why."

"He hasn't done his bad guy monologue, revealing his major plan."

That would have been useful.

"Can you save us?" Her scared voice now hopeful. She sounded full of life, not yet rotting in her living flesh.

"I'm sorry. It takes too much power and energy to operate the body. The more you can tell me about your captor and where we are, the more it will help."

A man spoke, his voice hoarse. "He said 'suffer my angels, it strengthens your energy.' "

This wasn't a serial killer with a unique signature. This was dark magic.

The door opened and the hooded figure strode in. Fear crackled through the atmosphere, poisoning everyone. Everyone fell silent except for laboured breathing from those close to death. He set a ladder beside my cage, climbed up and unhooked it.

"Another one who couldn't wait until the ceremony. You disappointed me."

As he lowered the cage, our eyes met.

I sat bolt upright, gasping. Sweat shrouded my body. My breath was entombed in my chest, as though part of me was trapped inside the body. Hijacking the dead was dangerous. Inexperienced necromancers sometimes remained trapped inside the corpse until they too, withered and died, leaving their empty bodies decaying in their beds.

I rushed to the bathroom and switched the shower on. As icy water pummelled my skin, I sank to the bottom of the bath, cleansing myself with a salt, olive oil and rosemary leaf paste. The light switched on and Saffron stood in the bathroom, looking like she'd been dragged through thorns.

"Either you didn't 'meditate' properly, or someone's been having raunchy dreams."

I hugged my knees, my teeth chattering. "I was in a body. In the cage. I saw the killer. And he saw me."

"Shit."

She switched off the shower and threw a towel around my shoulders. I pulled it around me, my trembling fingers struggling to hold it.

"Can he trace you?"

"One victim said he wanted energy. The more they suffered, the more energy they produced. The killer mentioned a ceremony."

"That sounds like a summoning ritual. Be careful, Indy."

"He's no match for Necro Girl." I tried smiling but couldn't even convince myself it was genuine.

"He's not a rebellious teen experimenting with dark magic to seem cool and daring. He's torturing people to death to summon something." She pulled me out of the bath and dried me. Normally that kind of intimacy made me uncomfortable but part of my mind was still in that gibbet cage. I needed to be touched, to know I was alive. "Whatever he's trying to summon is powerful."

"He may be powerful but he's targeted 'normal' humans. Not a necromancer who's spent her life studying the black arts."

"Look what effect being in the body has had. Next time you go for a spiritual jaunt, tell me, so I can guard your body. I don't want to find some evil entity has hijacked you. It'd break my heart to decapitate you and take ownership of your dragon collection."

"You're not having my dragon collection."

"Dead girls have no need for trinkets."

I returned to my room and dressed. I might as well use my ruined night to research. Immersing myself in someone else's nightmares might protect me from my own.

Saffron sighed. "What are we looking for?"

"Whichever summoning spell requires starving people to death. And how to stop it."

My phone rang. Saffron and I exchanged glances.

Jax's voice sound strained. "There's another one."

* * *

The woods were dark. Cold. Wind whispered secrets through the trees. Leaves and sticks crunched beneath my feet as though I walked on the bones of the dead. Floodlights turned trees into demons with gnarled hands designed for ripping out hearts.

A person-shaped gibbet cage swung from a branch with an almost skeletonised woman inside. She looked more like a creepy horror film prop than a human. The summoner wasn't just taking their energy and their lives, he was taking their humanity.

Jaxon approached. "Can you get more information from this one?"

The summoner saw me, maybe sensed my energy. He could track me. If he had me captive, he wouldn't need the others. My energy could raise whatever he was after, then he would kill them.

"Yes."

He smiled and handed me the protective gear. Every instinct warned me against this. I couldn't let anyone else die. It could take the police weeks to learn what I could learn in one touch. The price I would pay was worth it.

After donning the gear, I arranged my black candles and wrote on the leaf.

"Not more witchcraft shit," an officer muttered.

I lit the candles, biting my tongue. If I touched him while I touched the corpse, he would see what I saw, feel what I felt. As tempting as it was, I knew how to come out of it. He didn't.

I opened the cage, slipped the leaf in her mouth, held my wand and put my hand on the woman's face.

"*Mortuis resurgemus.*"

Pain pulsed through my head, blood trickling from my nose. The woman's limbs twitched as they regained muscle and fat. Her face changed from a wizened corpse to a plain woman in her twenties. Brown eyes blinked.

"I'm Indiana. Show me how you suffered."

So much pain. Unbearable hunger and thirst. Overwhelming, nerve-shredding fear. Darkness. Voices. Cries. A man in a hooded robe. Then I felt his energy. Dark, electric, dangerous.

"Cry, scream, fight. I need your suffering. Let me feast on your torment."

The woman screamed and thrashed. The cage swung. I struggled to keep contact with her.

"I'll kill you!"

The officers fled.

"What's your name?"

"Lydia Milton."

"I'll avenge you, Lydia."

I withdrew my hand. Light faded in Lydia's eyes. A tormented cry escaped her lips. Her body shrivelled as it died a second death. I dropped to my knees, fighting to breathe. I clawed the mud, craving the energy the earth provided. Black spots danced in my vision and agony ripped through my stomach. Blood dripped from my nose. Pain screamed in my head. Leaves beneath my hands rotted until I knelt in a ring of dead foliage.

I ran.

I stopped when I was away from the crime scene, sat against a tree and used my claw ring to tear my mental agony out through my skin. Pain took my focus away from their suffering, each slice waking me from the nightmare.

Blood dripped off my fingertip onto a leaf. Footsteps approached behind me. I pulled my sleeve over my cut and jammed my hands between my knees.

Jaxon crouched beside me. "I've never seen you kill anything. Do you have Midas powers?"

"Grounding myself in earth dispels the power, but his power is wrong and when I dispelled it, it killed. Luckily it was only greenery. Had someone touched me..."

Jaxon leaned away. "Did you learn anything?"

"The room looks like it's underground. They were yelling, so wherever they are, nobody's heard them. He's collecting their energy to use in a summoning. Suffering, despair, terror or grief are powerful."

"What does he want to summon?"

"I don't know. But we have to stop him before he achieves it."

"Why do they scream when you return them to death?"

"Because it hurts."

"Them or you?"

"Both."

He offered his hand then withdrew it slightly before extending it again.

"I won't kill you." I stood without taking his hand.

"Sorry, but I don't know about this... stuff. You killed those leaves."

"They were already dead. I sped up their decomposition."

He stared at his feet, looking awkward and afraid. I didn't want him to be afraid of me.

"Let me know how it goes with Lydia's family."

"I'll have to confirm her identity."

"Fine, don't believe a dead woman."

"I do. But my boss likes procedure and corpses aren't 'reliable witnesses'."

I returned to my hearse and slumped in the driver's seat. Dawn's pale fingers clawed though night's flesh as I drove home.

Saffron was using my laptop when I entered. "I found his potential ritual. 'Starve to feed'. He's starved many people, so I dread to think what he's feeding."

I recounted what happened then went down to the shop to collect ingredients for a tracking spell.

"How will you find him when you don't have anything of his?" Saffron asked as I arranged everything over a map.

"I'm hoping his victims' anguish will lead me to him. He needs their pain to strengthen his spell. That will be his undoing."

I concentrated on his face, his victims' faces, the underground chamber and focused my will into the crystal I dangled over the map. Images flashed through my mind, the torment, the fear, each one growing stronger until their agony was mine. I dropped the crystal. It landed on a patch of green.

"Someone would notice gibbeted people in a field. They're underground. I *saw* them."

Saffron brought up Google Maps and zoomed in on the field. A lone building occupied the spot.

"It's a church." She frowned. "Your nose is bleeding."

I touched it. Blood glistened on my fingertips. I wiped it then phoned Jaxon.

"My boss will kill me if I drag officers out to a church because a crystal landed on a map."

"You're right. It's probably a waste of time." I hung up.

Saffron narrowed her eyes. "You're going."

"I can't risk him completing the summoning."

"I'm coming."

"We need to open the shop in a few hours."

"Selling incense sticks won't matter when we're strung up in cages cursing ourselves for not bringing snacks to slow our decay."

After cleansing and casting protection spells, we climbed into my hearse. Saffron navigated through country lanes, surrounded by woodlands. I parked outside rusty church gates. Ivy strangled the bars. The graveyard was overgrown, nature reclaiming the dead and the stones that guarded them.

I must have been off my game when we were looking at the map, because now I realised where we were.

"This isn't good," I said. "This is where that vampire that makes Dracula look like a *Twilight* wannabe rests."

Saffron snorted. "Every teenage occultist has tried to raise Adriel."

"None starved people to death to feed him."

I took items from the hearse then we scaled the gate. We weren't exactly dressed for covert surveillance and rescue. Saffron wore a red latex dress, whalenet tights and chunky boots. I wore a pinstripe mullet skirt, a mesh skeleton print top and a pirate jacket. But we could die tonight and we didn't want our ghosts to be stuck dressed as burglars.

I could feel someone watching us, as though the dead were about to rise and claim our bodies for their own.

Saffron pushed the door. Locked.

"Of course a crazed necromancer would lock the door," I muttered.

Stained glass saints judged us as we prowled beneath them. I found a small ground level window and kicked it. Spiderweb fractures crawled over the glass before it surrendered.

Saffron raised her eyebrows. "Why not yell 'hey murderous summoner, we're outside'!"

"And ruin the element of surprise?" I switched my phone torch on. "Clear."

I shimmied through feet first, my boobs wedging before I slithered free. Saffron got stuck. I giggled as I pulled her free. We were definitely not stealthy superhero types.

Broken pews, old bibles and a rotting altar littered the small room. Dusty robes hung against a door as though a priest hanged himself long ago and was waiting for his corpse to be discovered.

We entered the main part of the church. I expected to see gibbet cages lining the aisle, their victims queuing for their unholy union with Death.

"There must be another cellar," I whispered. "Fan out."

"Have horror films taught you *nothing*? That's the quickest way to get a skewer through the boobs."

"If I see a skewer heading for your boobs, I'll tackle you."

Saffron poked her tongue out and headed right. I walked down the aisle. How many couples and coffins had made the same journey? I stopped by the altar. Stone slabs were laid in the floor, some epitaphs still visible. Around one slab, scrapes and stone dust covered the floor tiles. I waved Saffron over.

"Creepy guy's hiding in a burial vault. Should've guessed," she said. "Bets on there being a convenient crowbar around?"

"We can raise the dead; raising a stone shouldn't be difficult." I pulled a crowbar from my jacket. "Is this convenient enough for you?"

We moved the stone aside. Stairs led down into the darkness. I shone my torch down then descended.

Pungent odours of death, decay and bodily waste perfumed the air. I gagged, covering my nose and mouth. Walls to the neighbouring crypts had been knocked down into one large room. Lanterns provided a fiery glow, casting twisted shadows on the walls. Gibbet cages filled the space. Rotting people slumped inside them. Wet patches stained the floor. I could taste death.

Swallowing my vomit, I lit my black candles, opened the nearest cage, and placed my hand on the corpse's head, my wand clutched in my other hand.

"*Mortuis resurgemus.*"

The corpse returned to life.

"Where is he?"

"I don't know."

I coughed, trying not to inhale death's putrid fragrance.

"Are we waking them all?" Saffron asked. "That's too much backlash."

"We need to know what he's planning."

"I wanted an excuse for an orgy."

We resurrected every corpse. So much energy surged through me, I could power an entire city.

One corpse whispered "mausoleum."

We left the church for the furthest corner of the graveyard where a lone mausoleum lurked. Power emanating from it made me feel sick. Saffron clutched her stomach. Wind whipped us, bending trees to its will. Electricity sparked in the air.

I kicked open the door. Dagan the Undead stood beside an open sarcophagus which contained what looked like a recently deceased corpse. Except this one died a thousand years ago. Dagan was reciting from a book, pouring blood into the corpse's mouth.

I couldn't see any of his groupies in the shadows, so maybe we still had a chance.

"Dagan, don't!"

I ran around the sarcophagus, my crowbar raised. Saffron crept around the other side, hopefully as yet unseen.

"*Mortuis resurgemus*!" he cried out.

"No!"

We were too late. Before we could act, Adriel was on his feet, staring wildly round. He didn't wait for polite introductions. He leapt at Saffron.

I scrambled over the sarcophagus and stabbed him with my crowbar. He turned, bared his fangs, then fled; Saffron collapsed, blood dripping from her neck.

And I still had to tackle Dagan. I slammed him against the stone wall, trying to pin him.

"Once I've drunk his blood," he gasped, "I'll become immortal. You cannot stop me."

"You have to find him again first."

"Use black magic against me, and you'll be damned." He smiled.

"Shame you won't be alive to enjoy it. Let their deaths be visited upon your soul."

"Indy, no!" Saffron called. "Remember our pact!"

"Some things are made to be broken. *Et punire mortuis*!"

The dead will punish.

I gripped Dagan's head, unleashing the aftermath from the corpses. He screamed as their suffering, pain, fear, despair, poured into him, fuelled with my rage. Blood streamed from my eyes and nose, while he withered, his skin shrivelling as he starved to death. Still he screamed.

Smoke rose from his head. I peeled my hand away, my skin blistered and raw. I hurried to Saffron and tied my jacket around her neck. My body shook as it weakened.

"You destroyed him with necromancy," Saffron murmured. "This will be revisited on you threefold."

"I had to stop him."

"I feel faint."

"If you die, I *will* resurrect you as my zombie slave."

Saffron managed a weak laugh. "What do we do now?"

A good question, with only one answer.

"It's time to kill the dead."

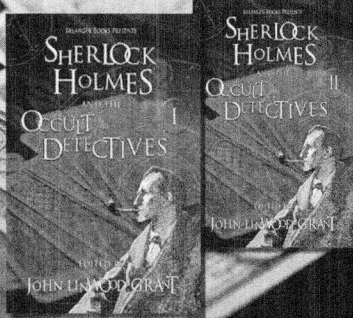

THE HAND THAT SHUTS EVERY DOOR

JONATHON MAST

Miller prodded the mound of quilts. He hoped it was a person under there. He kept his eyes darting from the pile to the door and back again.

The form under the blankets awoke with a start. "I'm not crying. I'm not crying." Layers of dirty fabric fell away to reveal the head of a young man, maybe sixteen. His face bore no stubble, but his voice was deep. His mouth was the right size, and his eyes weren't made of wax. Those were good signs.

Now that he knew it was probably safe to look away, Miller watched the door. "It's time to go."

The kid looked up, hope in his gray eyes. "Who are you?"

"Miller. Come on." He glanced around the room, his eyes inspecting each corner, always returning to the single doorway.

The space was small, perhaps three meters by three meters. Faux wood paneling hid the concrete walls. The ceiling was low, two and a half meters, and of yellowed paneling. Dirty brown carpeting covered the floor. Only a dingy twin mattress – no frame – furnished the room. Cold light streamed in from a narrow window near the ceiling. Humidity thickened the air.

The boy struggled to his feet, untangling himself from the blankets. His clothes were filthy. He was tall, even taller than Miller. He didn't look to have been sweating, even under the blankets. "You can leave?" he asked.

"That's the plan." Survey complete, Miller's eyes locked on the doorless entry. He had heard nothing since descending to the basement. The House didn't even creak above them, and he knew he had left his two companions on the main floor. Something was off about the place, and he intended to leave without incident. Finding the kid gave him an excuse to turn back now.

"You can open doors?"

Miller's attention snapped to the kid. "What's your name?"

"Brian."

"Well, Brian, it's time to go."

He stepped into the hallway, and looked back. A window above the opening showed gray sky. It should have showed the room he had just left.

He turned to gaze deeper into the basement, down the hallway toward what he guessed would be two or three more rooms, though he did not go in

205

that direction. Not yet. He had already passed through a series of chambers, all faux-wood paneling, all brown carpet, all like the room where he found Brian, but with a separate way in and way out.

Miller had no idea who Brian was, but the kid didn't feel like part of the basement. He felt as if he had been caught here. It was good Miller came when he did; people usually didn't survive in places like this too long. Not that Miller was somehow more badass than anyone else; he just knew when to get out.

Brian stepped behind him, facing the same direction. "When I couldn't open the door to go back, I stayed here. I didn't want to go any farther. That's where it lives."

"You ever see it?"

"I think the hand belongs to it." His voice stumbled.

"The hand?"

"The hand that shuts every door."

"There were no doors on my way here." Miller glanced back, the way he had come.

Brian whispered, "No. There weren't."

"All right. Let's get outta here." He turned to leave.

At the far end of the hallway was a door. A door that had not been there when he first looked. Miller swore and reached for his cellphone.

He had no service. Of course he had no service. What kind of House would this be if he had service while near a high danger area, with a kid in tow?

"How long you been down here, Brian?"

"I don't know. Two, three days?"

Miller replaced the cell and examined the door. It was constructed of dark wood and rose to about thirty centimeters from the ceiling, about a meter and a half wide. A plain brass doorknob was affixed to the left side of the door. He could identify no hinges. "What have you been eating?"

"Nothing."

"You drink anything?"

"... no."

"That's not good." Miller's right hand hovered near the door.

"I know. But I'm not thirsty."

"That's very not good." He let his fingertips brush the surface. Nothing happened. "What brought you here?"

"I live about a mile down. After the fire, I wanted to see if there was anything left worth taking. Wandered down the basement steps. Been here ever since."

Miller resisted the urge to curse. The last recorded fire in the place was a little more than two decades ago. The kid wasn't a norm, then. Best not to tell him; he probably didn't realize that he'd been dead for a while.

Miller's fingers traced a light path to the doorknob and gripped it. It turned easily. He glanced back at Brian without pulling at the knob.

The kid shook his head. "It always lets you turn the knob. It'll let you open it, sometimes, too. But it never lets you out."

Miller blinked, faced the door, and opened it gently.

The door swung open freely, and they could see the room beyond, a perfect copy of Brian's, except for the closed door on the far end.

The knob tore from his hand. Miller had the briefest impression of a hand shoving the door shut. It happened so fast he couldn't tell any details except that maybe, maybe, the hand was Caucasian.

Miller heard the sound of countless more doors slamming shut. He felt a warm push of air from behind him, from deep in the basement. He spun, but nothing appeared out of place.

"It's like that every time. I gave up and slept." Brian shrugged.

Miller wasn't surprised. Door-based images like these were common. The problem wasn't the doors, nor the kid. The problem was whatever lay in the last room. Whatever was using the hallway to breathe.

He closed his eyes and reached for the knob with his left hand, keeping his eyes closed. He opened the door just a crack, while raising his right hand, palm open. The door seemed to shiver ever so slightly, as if awaiting some ecstasy. The hand came. Miller snatched at it and missed. The door slammed shut again, and a cacophony of echoes surrounded him.

Miller turned to face Brian. "You haven't gone any further?"

"No."

"Did you try the windows?"

"No. They're too small for me to get out. Look at them!"

Miller turned his attention to the window. The light outside was still bright, though the sky was marbled gray. The glass was thick, but clear. He retrieved a small hammer from his pocket. It was always useful to be able to pound something.

"Let's go back to your room, Brian. I don't like the hallway."

When they entered, Miller took the mattress and lifted it with a grunt, bracing it against the wall under the window. Once there, he climbed it so his face was level with the glass.

"What's happening?" The kid's voice shook with uncertainty.

Miller contemplated the glass as he talked. "There are things in the world, Brian. Old things." He began tapping around the frame, trying to determine if he could remove the whole thing. "They're sleeping now, which is good. But every once in a while they roll over, and when that happens, strange stuff seeps out from their dreams. My friends and I travel to wherever something strange happens and investigate it. Make sure it's not a hoax. If it is a hoax, which is most of the time, we debunk it. If it's real, we make sure no one disturbs the sleeping one. Because we don't want to wake any of them up. Not even one."

"So you're here to fight it?"

Miller resisted the urge to laugh. "No. You don't fight the old ones. You leave them be and hope they don't notice you. We're nuts enough to do as much as we do." The sealant was old. Taking the claw of the hammer, he began prying some of the old rubber-like material out.

Miller continued, "When we investigated the fire, we figured it might be an old one. You heard what happened?"

"Yeah. Mrs Friedrichs burned the place down. Killed her whole family. The paper said she went nuts."

"Doctors say it was postpartum depression. We dug a little deeper, though. Apparently she starved her infant, refused to give it any milk. Even after it died, though, she would carry it around, put it in the high chair, all sorts of creepy stuff. Family didn't even notice; a guest did. Called the cops."

Brian looked disturbed. Miller chose not to notice. "Apparently Mrs Friedrichs was serving breakfast the day the cops were supposed to come, and poured gasoline on the cereal instead of milk. Something must have been wrong with the family, too. They ate it right up. Then she lit the place on fire and died in the blaze."

More sealant came out, along with a lot of dirt. Miller noticed that there were no webs. Whatever was down here was creepy enough spiders didn't want to be near it.

Brian changed the topic. "Why does it shut doors?"

Miller grunted a quiet laugh. "Best not try to understand. Their thoughts are so alien to ours, be happy you got a door and a breathing hallway. Could've been far worse. Like whatever happened to Mrs Friedrichs."

Finally enough sealant came away. He grabbed hold of the frame and pulled, grunting. It budged a fraction of an inch, then an inch more. Finally the whole frame started sliding out.

"You think you could fit out this hole?"

"I don't know. I'll try." Quiet excitement crept into Brian's voice.

Milled stepped off the mattress and let the kid scurry up. He plunged into the window. He grunted, pushing with all his might. Miller came up behind and braced the kid's feet, trying to help.

Brian's slender shoulders allowed him struggle out. He scurried around and peeked back at Miller. "Are you coming, too?"

"No. I won't be able to fit. But go to the main floor. Don't come back into the basement. I have friends there, other people I work with. Tell them I'm in the basement, but that they shouldn't come after me. Either I'll find a way out or they'll have to leave me." Miller kept his voice even. Maybe the kid would be able to get the message to them, maybe not. Either way, he felt better treating Brian as if he was alive.

"Wait – can't they just get something to widen this hole?"

"They can try, but it's not likely to work. That kind of thing hardly ever works."

Brian locked eyes with Miller. "Thank you," he whispered, and was gone.

Good. The kid was wherever he was meant to be, not trapped in some dream. Miller returned to the hallway. Time to get to work.

He opened the door and let it slam in his face. His eyes slid over the image. He couldn't focus on it, but it was a hand. Disembodied, definitely Caucasian, no distinguishing marks. Severed cleanly at the wrist.

Once again the hallway breathed on him, sighing. He did not acknowledge it.

If he got the hand out of the way, he might be able to open the door and get out of the House. The only question was whether or not he could make contact with the apparition. If it was solid, he'd be able to do something.

Miller reached for the knob with his left hand again, preparing his right to swipe. Every time the hand appeared, it had made contact with the door at the same place. He would be ready.

His fingers wrapped around the knob, one by one. The metal felt cold against his skin. He breathed in once. Twice. He rotated the knob and let the door fall open.

There. The hand. He grabbed at it.

The hand grabbed back.

Title: ANGEL'S INFERNO
Author: William Hjortsberg
Publisher: No Exit Press
Format: paperback / ebook
Reviewer: William Meikle

No more an angel but still a favourite.

I've been waiting for this for decades, the posthumously published sequel to *Falling Angel*. I read the latter even before seeing the seminal *Angel Heart* movie, but both of them have, over time, become interwoven and hard to untangle in my mind.

Reading this book, I have a feeling the same thing happened to the writer; even more than *Falling Angel*, this sequel is intensely visual, even if it does at times read like an extended travelogue for Paris.

Harry is no longer quite Harry, Johnny Favorite having come to the fore. And Johnny has expensive tastes that require funding. *Angel's Inferno* starts with this emergence, taking up directly at the end of the first book, and in its early stages maintains that book's almost Hammettesque clipped diction, but by the time Johnny has fully re-emerged we're into more Chandleresque territory, of posher hotels, better food, richer dames and higher stakes.

Johnny is on the trail of Louis Ciphre, but the enigmatic magician isn't showing himself. Instead, Johnny is led along numerous blind alleys and sidetracks, in posh restaurants, smoky Jazz clubs and cabaret joints and on to chance encounters with several of Paris' more notable literary figures. He's

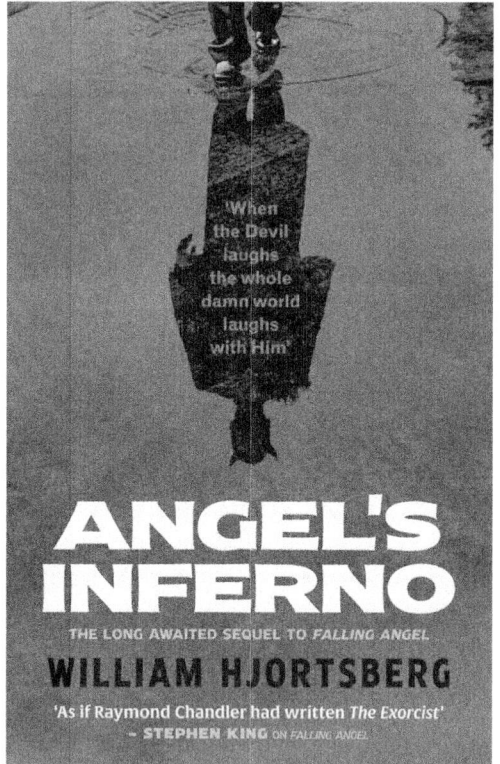

also got cops on his trail, having fled from the carnage at the climax of the previous book. What follows is the aforesaid travelogue through Paris, during which Johnny discovers more about the Satan worshipping cult he's embroiled in, and old memories return to fill in the blanks.

As usual, there's a MacGuffin driving the plot forwards through the middle section. Whereas in *Falling Angel* it was Harry chasing himself that was the driver, here we have a tale of Judas Iscariot, thirty pieces of silver, and thirty members of a cult – not just any old Satanic Conspiracy, but one dating back to the Crucifixion and the birth of Christianity. This leads Johnny to a secret society meeting in the vaults of the Vatican, but Louis Ciphre is still nowhere to be found.

The book turns on a pivot when Johnny and Louis Ciphre finally meet in a quiet Parisian park. It's a remarkably well done scene that harks back to the first book while setting the tone for this book's final act.

With the police net closing in on him, Johnny must resort to desperate measures and sacrifices must be made. Infiltrating the secret society in the Vatican he returns there for one final confrontation with Louis Ciphre, and the culmination of all that has gone before.

I saw the ending coming some way off, it's inevitability being obvious even though Johnny, like Harry before him, doesn't see it coming.

I enjoyed this book, but didn't love it in the way I do with *Falling Angel*. I think it's the fact that Johnny is far less sympathetic a character than Harry Angel, and that dilutes the tragedies that unfold as it's that much harder to root for him. That said, the writing is as crisp as before, and I enjoyed the journey. It's just not the classic I was hoping for.

But in comparison to *Falling Angel*, what is?

My rating? 4/5 satanic pentacles.

Title: A HOLE IN THE WORLD
Author: Weston Ochse
Publisher: Solaris
Format: paperback / ebook
Reviewer: Dave Brzeski

This one is a little hard to classify, The FantasticFiction website claims that it's urban fantasy, which considering it's set in a world much like ours, but in which the creatures of myth and fairy tales actually exist, is understandable. But this is no modern retelling of any classic fairy tale, and despite having an attractive, young female lead character, neither is it paranormal romance.

Imagine if Frederick Forsythe had dryads, gnomes and pixies in his novels, and you'll get the idea. This is as much thriller as it is fantasy.

Laurie May, codename 'Preacher's Daughter', is a soldier in the employ of Special Unit 77. I was instantly reminded of an obscure TV show from a couple of decades back called *Special Unit 2*, and indeed, there are a few similarities. Special Unit 2 were a secret Chicago P.D. unit who solved violent crimes committed by mythological creatures: aka 'Links', while trying to hide the Links' existence from the general public. Special Unit 77, are a special ops. organization whose remit is to protect the United States, and her technology from supernatural exploitation by other countries – so, similar, but rather bigger in scope. I contacted the author to ask him if that old, short-lived TV show had been an influence, but he'd never even heard of it.

Iron Hat, a town on the First Nation reservation in South Dakota has 'disappeared' – in the sense that it has not only gone, but been wiped from the memories of all but one man. For example, every time someone tries to drive to the town, they find themselves turning back, having forgotten it was ever there. That one man, Francis Scott Key Catches, remembers the place is due to his narcolepsy having put him to sleep when whatever happened... happened.

Not only is evidence of the existence of Iron Hat disappearing as fast as it can be uncovered – if this wasn't enough to interest Special Unit 77 – a similar thing has happened to Graves Hill, a town near Leicester, in the UK. So it is that Preacher's Daughter finds herself part of an exchange of team members with their British equivalent, The Black Dragoons, to help facilitate them working together on the case.

There have been several incidents of large groups of Fae invading the human world – leaving their own reservations, so to speak. There's a quite obvious correlation between the Fae and the First Nation tribes in that they were pushed out of the lands they once ruled, and in the way they are treated by the new incumbents. It seems that these Fae may be running from something that is even older.

Preacher's Daughter, and her new team-mates have to try to form some sort of alliance between themselves and the Fae, but the Dryads, who used to keep the dimensional barriers secure, are also disappearing.

Communication between Special Unit 77, the Black Dragoons, their various governments and the Seelie, and Unseelie Courts is, to say the least, fraught. There's little trust to be found, and everyone has their own agenda.

It's a fascinating, well-paced novel which left me wanting more. As it happens, there is more, and not just in the way of future books in this series. Preacher's Daughter has previously appeared as a secondary character in the two volume Sky series (*Burning Sky* & *Dead Sky*) wherein some of the previous events referred to in this book occurred. There's also a four book related series involving SEAL Team 666, who are mentioned in passing here.

Title: RAINBRINGER: ZORA NEALE HURSTON AGAINST THE LOVECRAFTIAN MYTHOS
Author: Edward M. Erdelac
Publisher: Self-published
Format: paperback / Kindle
Reviewer: Rebecca Buchanan

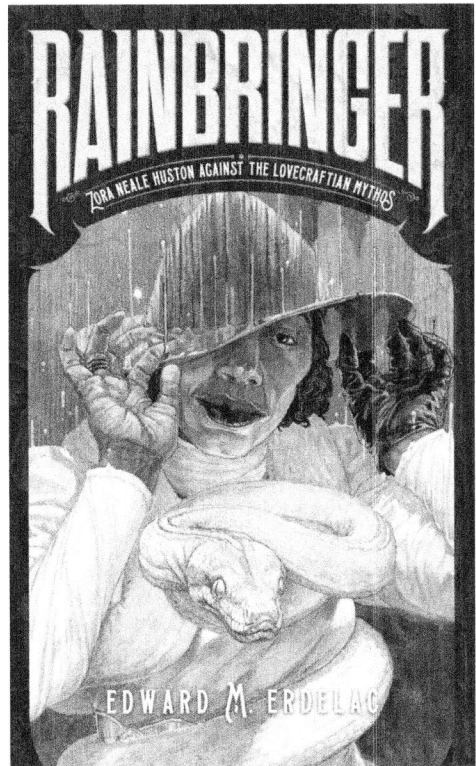

There are things far older than humanity, or even humanity's Gods. Some of them sleep. Some of them are beginning to wake, their servants performing perverse rites and sacrifices in their honor. But there are also those who stand guard against these ancient evils. One such guardian is known as Rainbringer...

Zora Neale Hurston. Now, she is a giant of the literary canon, mandatory reading in schools across the United States, in classes covering everything from anthropology to history, to African-American Studies, to religious studies to literature. During her lifetime, however, her fame and fortune fluctuated wildly, sending her from the streets of Harlem to the ivy-covered walls of academia to the bayous of the deep south, and back again. In Erdelac's skilled hands, Hurston's writings, wanderings, and dramatic life create the frame around which he builds his fictional collection of adventures. Each of the eight stories in the collection – beginning in 1925 and ending(?) in 1975 –   slowly build on one another, increasing in trepidation and terror. Over the course of decades, Hurston is initiated into Voodoo, earns the name of Rainbringer, comes to the aid of ghosts and fallen priests, battles madmen and monsters, collects ancient songs and words of power, and finally – well, that would spoil it.

I became a fan of Erdelac after reading his novella '*Mindbreaker*' in *Bond Unknown*. *Rainbringer* just reinforces my admiration. It is a fantastic weaving of the real and the horrific, lyrical and mystical, but also profoundly political and deeply angry. As an intelligent, well-educated, and strong-willed African-American woman, the real Hurston repeatedly encountered and faced down sexism and racism (sometimes blatant, sometimes couched as benevolent patronage); here, she faces such real-life horrors in addition to the horrors of the Old Gods and monsters from the stars. Sometimes, the two overlap, with misguided or power-hungry humans attempting to claim the power of the dreaded Old Ones for themselves (it never ends well). Sometimes she has allies, other times she is alone, with only her own strength and determination to hold her steady in the face of such terrible power. Through it all, she maintains her sense of humor and her sense of her own worth.

Rainbringer is unique and compelling, a formidable addition to the weird fiction canon. Highly recommended.

Title: MAJOR CRADDOCK INVESTIGATES: FOUR VICTORIAN CHILLERS
Author: Paul Finch

Publisher: Brentwood Press
Format: Kindle
Reviewer: Dave Brzeski

In the North of England in the 1860s, police detective Jim Craddock has his work cut out in the teeming slums and smoggy backstreets of the Industrial Revolution – without finding himself confronted again and again by supernatural villainy...

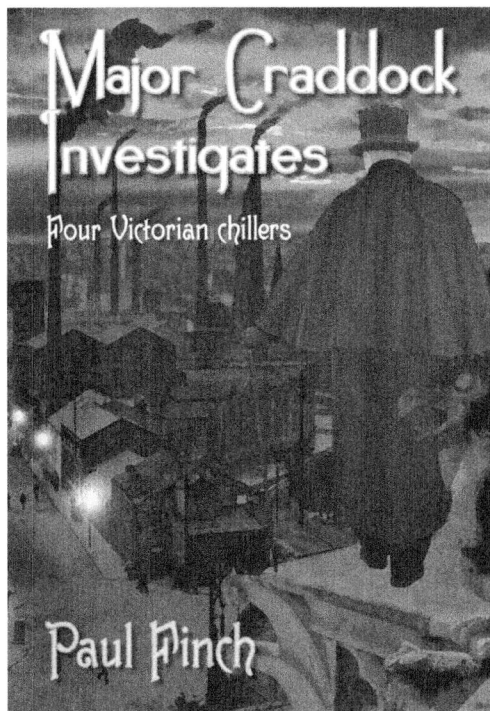

Major Craddock Investigates

Four Victorian chillers

Paul Finch

'It's grim up North', or so the popular cliché goes. Clichés tend to have some basis in truth, though, and it can't be denied that it was pretty grim in Wigan in the 1860s. There were many theories in the military as to why Major Craddock resigned his commission to take a job in the newly formed civilian police force, but the truth of the matter was that he felt he could do some good.

The collection opens with, 'The Magic Lantern Show'. There had been a brutal murder. To be honest, even amongst the large ex-patriot Irish community of Wigan, James O'Hare was not going to be mourned. It was the nature of the murder that caused Craddock so much concern. That such a strong, powerful man, even in his cups, could be strangled so brutally that his larynx was crushed and his vertebrae cracked was astonishing. It's only after a further two similar murders, both of women, that Sergeant Rafferty can be persuaded to share a theory that he'd been forcefully dismissing.

The historical background of these stories is as well-researched as I'd expect from Finch, and there's also a little social commentary on the subject of the Irish who came to England in search of respite from the aftermath of the infamous potato famine.

'Shadows in the Rafters'[1] involves a decommissioned coal mine, and disappearing children. A man has been arrested for holding up a coach occupied by the Reverend Pettigrew, and his brute of a Boer servant, Krueger. The police are already persona non grata amongst the mine workers and their families. This is due to their part in keeping the peace when Messrs Byrtle & Son closed the mine with little warning, nor recompense to those who relied upon it for their daily sustenance, because new safety regulations

---

[1] A newly revised version of this story has been generously donated by Paul Finch for inclusion in the promotional Occult Detective Magazine #0.

made it too expensive a proposition. So Craddock is unaware of the rumour about the mine... the screams in the night and the missing children. He learns that Kreuger guards the mine and forcibly keeps people away from it, suggesting that his employer has something in there he needs to hide.

'The Weeping in the Witch Hours' is the longest tale in this collection, and for me the best. It's unusual in that Major Craddock is offered a chance at taking a well-earned holiday – albeit in the Fens, in February! It's soon revealed that Chief Justice Reginald Bowery wants him, on the request of his friend the Bishop of Peterborough, to investigate a case which they feel requires a certain expertise. This is the first time we've seen Craddock take on a case because it may involve the supernatural, as opposed to simply encountering it in the line of his normal duty.

Of course, there's no proof of anything unnatural in the case of the rector of St Brae's Cathedral, and his deacon both being found dead on the altar, with no mark on their bodies other than a look of abject terror. This could be natural causes, but if it was murder, it could be either by the hand of earthly criminals, or some occult agency... or both!

I recently reviewed three collections of Finch's shorter work for John Linwood Grant's *greydogtales* blog.[2] I commented there on the author's habit of introducing secondary characters the reader might sincerely hope to see again. The aged, but by no means infirm, white witch of King's Fen, Madam Godhigfu can certainly be added to this list.

The final tale, 'The Coils Unseen' is brand new to this collection. Major Craddock has trailed the infamous George Krugg to an abandoned prison ship, where he finds the local garrison of Hussars, local part-timers, to be a cowardly lot who would not be persuaded to board the apparently 'haunted' hulk.

One man agrees to accompany them however, so Craddock, his right hand man Munro, young Constable Palmer and Corporal Kenton climb on board the huge, dark, forbidding prison ship. I was unaware of the history of such things before reading this story, but it seems these ships were quite common in the days before prison reform, and they made the purpose built prisons seem like luxury hotels by comparison.

It turns out that Krugg had a reason for luring Craddock on board the ship – there was something he needed Craddock to see for himself... something that brought the very nature of evil into question. It's a fascinating idea, and I will be intrigued to see how this knowledge might impact on Craddock's future cases.

[2] http://greydogtales.com/blog/paul-finch-ill-met-by-darkness/

Title: THE GIRL WITH GHOST EYES (The Daoshi Chronicles Book One)
Author: M.H. Boroson

Publisher: Talos
Format: Paperback / ebook
Reviewer: Rebecca Buchanan

"Martial arts and Asian magic set in Old San Francisco... A fresh take on urban fantasy that kept me up late to finish."
— #1 *New York Times* bestselling author Patricia Briggs

M. H. BOROSON

THE GIRL WITH GHOST EYES

San Francisco's Chinatown at the end of the nineteenth century. Xian Li-lin is quietly attending to her duties at the Daoshan temple when Liu Qiang enters and requests her assistance: a deceased friend is trapped outside of the City of the Dead. The only way for him to enter is with a soul passport, which Li-lin reluctantly agrees to deliver. She crosses over to the spirit realm, only to discover that she has been deceived. If she can't find her way back to the human realm, her father will be murdered. But even if she does find her way back, greater dangers still lie in store for her: ruthless Chinese tongs, a hungry tiger spirit in human form, a one-armed sorcerer, a vicious spirit snake, a monster out of myth, and a magical conspiracy bent on destroying Chinatown...

Occult detective/urban fantasy stories which deal with non-Western mythology are a rare treat. When I came across *The Girl With Ghost Eyes*, I immediately added it to my wish list, but it was some time before I was able to actually purchase the book and sit down to read it. I'm sorry that it took me so long to do so: this is an excellent tale filled with compelling characters, intriguing magic, and a seriously dangerous villain.

Li-lin herself is deeply sympathetic and appealing. The daughter of one Daoshan master and the widow of another, she believes herself to be happy. She misses her husband terribly, and plans to remain a chaste widow for the rest of her life, but she also finds her work as a Second Ordination Daonu to be satisfying: she tends to the shrines of the dead, deals with the odd ghost, and occasionally assists her father in removing supernatural threats from the community. Then Liu Qiang comes to the temple, lies to her, traps her in the spirit world, and cuts a spell directly into her body, leaving her scarred for life. She has been betrayed and violated. Li-lin's anger at that violation grows, and in her anger she begins to discover who she truly is, what her abilities really are, and the lengths she will go to protect those she loves. Over the course of the book, she comes into her own, figuring out who she is outside of the men in her life. What starts out as a personal quest for justice evolves into a desperate attempt to discover the true extent of the threat against not just her immediate family, but her extended family: the people of Chinatown, of San Francisco, of California, and beyond.

Just as interesting is the cultural venue of *The Girl With Ghost Eyes*. I don't know of any other such stories which are set in a 19th century Chinatown, and which center the narrative on the immigrant Asian community; the only Caucasian characters to appear are a couple of policemen, a trolley driver, and some out-of-work rabble-rousers. While Boroson explains in his author notes that some of the religious and magical elements were tweaked for narrative clarity, by and large it is an authentic presentation of Daoist beliefs and practices. Li-lin's peach-wood sword is awesome, the names for the various martial arts positions are almost poetic, and the bagua mirror is fascinating. (Boroson also includes a recommended reading list for those who want to learn more).

The Girl With Ghost Eyes is a terrific start to a new occult detective series. The second book, *The Girl With No Face*, has now been released, and I can't wait to read it. Highly recommended to fans of *The Haunting of Tram Car 015* by P. Djèlí Clark, *Spectred Isle* by K.J. Charles, and *Servant of the Underworld* by Aliette de Bodard.

Title: FORGOTTEN HOMELAND / A MEETING OF TERRORS / COUNTING CROWS: ONE FOR MURDER
Author: Joe Talon
Publisher: Mirador Publishing
Format: paperback / Kindle [3]
Reviewer: Dave Brzeski

[3] The two novellas, *Forgotten Homeland* and *A Meeting of Terrors* are available in multiple ebook formats, but not in paperback. It may still be possible to find free download offers on those.

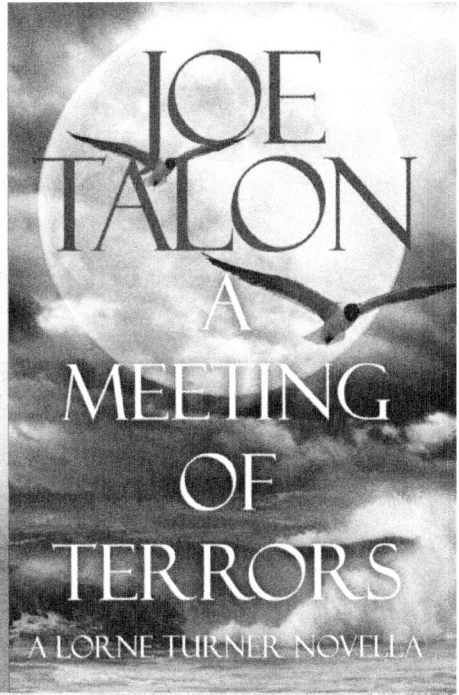

Every so often, I look through lists of free ebooks on Bookbub, or similar book promo sites, to see if anything strikes me as being potentially a good candidate to review in ODM.

Forgotten Homeland caught my eye mainly because it was a novella-length introduction to a proposed series.

Lorne Turner is an ex-soldier, with a pretty bad case of PTSD. He finds himself inheriting his deceased parents' farm, along with debts that outweigh the value of the

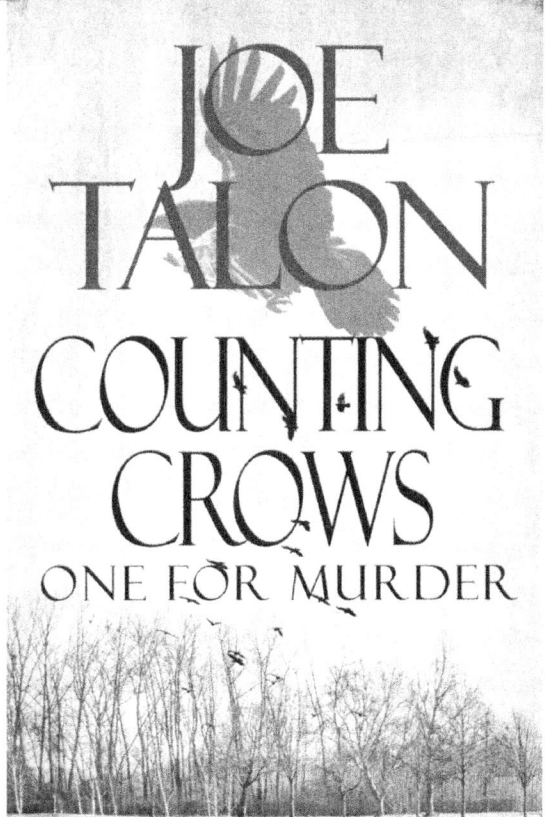

inheritance. His looks to be a life of nightmares, debt and unbelievably foul weather.

Having managed, with the help of his father's old friend and neighbour, to get things on an even keel, albeit temporarily, he finds his sleep disturbed by more than his oh-so-familiar nightmares. There's an old church on the property, which is only visited by the local cleric once per month; Lorne hears the banging of doors, and a woman's scream...

It's a fairly short tale, and the problem is quickly sorted, but as an introduction to Lorne Turner, and the reverend Thomas Hearn, it works well enough. It's well-written, and left me interested enough to immediately email Joe Talon to ask for a review copy of the first proper book in the series.

In the meantime, I read a second Turner novella – *A Meeting of Terrors* – in which Lorne is called in to assist with a nasty situation by a friend who knows he has the necessary skill-set to help. It seems a bunch of teenagers have got themselves stuck in a cave system, with the tide coming in fast. Lorne finds he's not the only person with climbing skills who was brought in to help – Ella Morgan is a local vicar, who also happens to be gay. It seems evident that the author had decided that the reverend Thomas Hearn needed to be replaced by a character with more potential for future stories. While I feel this was the right decision – Ella is a much better character – I liked Thomas Hearn too, and do hope that he won't just disappear from the series.

It will likely come as no surprise to our readership that the teens have disturbed something ancient, and evil, with which Lorne and Ella have to deal.

This is very much another short introductory episode, and while I enjoyed it, it's not absolutely necessary to read either of them before the first actual novel in the series.

Counting Crows: One For Murder hits the ground running. After yet another night of PTSD related nightmares, Lorne sets off for a run – a run which is interrupted by his finding the body of a young, evidently-murdered junkie. When they arrive, the police aren't much interested in his opinions on the ritualistic nature of the killing. Thankfully, before calling them, Lorne took photos of the body, with all those strange tattoos, which he soon shows to his vicar friend, Ella.

It seems at first that this could easily be an unremittingly dark book, and I've always been of the opinion that darkness needs some light to balance it out. Thankfully, Talon has such a talent for humour in his writing that I found myself laughing out loud on several occasions.

Having acquired one female friend, albeit one for whom he can't entertain notions of a sexual nature, Lorne, soon meets another woman of major importance in his life – Willow Hunter, a Wiccan, who runs the local magic shop. Much of the humour comes from the interaction between these three very different people. Talon, however, easily manages the most important task... he makes us like them.

Humour aside, things soon begin to get increasingly dark, as the power of the bad guys – and I'm not even talking about occult power here – starts to make life very difficult for them.

I will say no more about the story, except to mention that, while I admit to having no personal experience of veterans suffering from PTSD, and I have no idea if the author does, he handles that aspect extremely well – at least it was convincing to me. This is a truly gripping page turner of a novel that left me seriously looking forward to the next book.

Title: THE FALL OF THE HOUSE OF THOMAS WEIR
Author: Andrew Neil MacLeod

Publisher: Burning Chair Publishing
Format: Paperback / Kindle
Reviewer: Dave Brzeski

Assuming you're not one of those odd folk who skip to the back of a magazine to read the reviews first, you'll have already been introduced to Andrew Neil MacLeod's investigators of the weird and inexplicable, Johnson and Boswell, in the story, 'The Grey Men of Glamaig', in this very issue. While that story serves as a brief introduction to our heroes, I felt its greatest strength was in the tremendous potential it opened up for further adventures. At that time, I hadn't realised that this potential was about to be realised quite soon, and in a

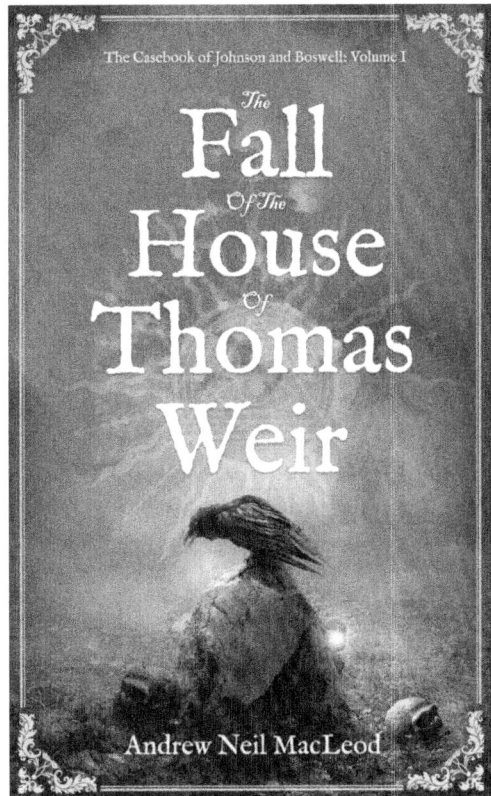

series of full-length novels.

Dr Samuel Johnson and his friend James Boswell are, of course, very well-known figures in British history, albeit Boswell is best known these days for his famous biography of his friend, Samuel Johnson. MacLeod's fictional tales of Johnson and Boswell take place when they were both living in Scotland, and detail their interest in things supernatural.

I had no idea on starting this novel just how far MacLeod was going to take our intrepid investigators. This is no simple haunting. Yes, it all starts out with some investigation of people going missing that doesn't look too taxing for them to explain… but that's just the beginning of a chain of events which could impact the entire World!

I have to be careful to not give away too many plot points here. Let's just say that the government's callous handling of a terrifying pandemic (who'd have guessed that could happen?) over a century earlier, had caused both the afflicted and the healthy to be sealed in the tunnels of Edinburgh Old Town, and a new city grew above it. It almost goes without saying that those buried streets and tunnels are not unoccupied, but the problem goes (literally) deeper than that.

In a story that quickly escalates, in both scope and body count, MacLeod takes us from long-lost civilizations all the way to cosmic territory. High points for me include one of the best descriptions of non-Euclidian geometry I've yet read, followed by a very Hodgsonian scene in which Johnson looks through a window and sees an entirely different world.

And I can't not mention that my favourite character in the book is Mrs Maggie Boswell — an absolute force of nature who could easily carry a book on her own.

Is it a perfect book, no. It's well-written, well-researched, albeit MacLeod admits to taking the occasional liberty with historical accuracy in his acknowledgements. I did find a couple of very minor anachronisms in the prose, but there is a fine line between slavish accuracy and a readable – to a modern audience – style. For the most part, I found it a laudable achievement for a first novel, and I will certainly be reading the follow-up volumes.

Title: THE DARK MAN: A PARANORMAL THRILLER
Author: Desmond Doane
Publisher: Self-published
Format: Paperback / Kindle
Reviewer: Dave Brzeski

Desmond Doane is a (not very secret) pseudonym for mystery and thriller author, Ernie Lindsey – the point being that all his paranormal investigator books are kept separate from the mainstream stuff.

This particular novel is book one of the *Graveyard: Classified* series, which involves the disgraced star of one of those paranormal ghost-hunter TV shows. Things went badly wrong on a special live broadcast episode of the top-rated show, and a five year old girl was hurt. The show, of course was cancelled, and Ford Atticus Ford's name was dragged through the mud. His partner and best friend, Mike Long, hadn't spoken to him since.

Now, although he'd made enough money to not actually need to work, he keeps his hand in as a consultant for any law enforcement officer desperate enough for a break in their case that they'd consider asking for his help, despite their not believing in all that supernatural nonsense. I have some sympathy for these non-believers. I also consider those TV shows to be nothing more than fakery, and the likes of Derek Acora to be nothing more than performers. However, what if they're genuine? Would they garner any more credence from the public, or more to the point, the authorities? The answer is, of course, no!

We learn more about the disaster that ended Ford's successful career, plus some details of an investigation he's on before Detective Thomas requests his help on a case. Thomas is investigating the murder of Louisa Craghorn. Complicating matters is the fact that her husband, Dave Craghorn, is suffering under what appears to be a very malignant haunting.

The story is told in that classic first person, noir narration style, so popularised by the great hard-boiled detective story authors of the past. We get a lot of that submitted to *Occult Detective Magazine*, and frankly some of it is pretty bad (though when it works, of course...). If I never read another opening where the knock on the detective's office door turns out to be from a beautiful redhead with legs that go on forever, it'll be too soon. Being *Occult Detective Magazine*, they usually turn out to be vampires. Thankfully, Doane doesn't go there at all. In fact he handles the narrative style to perfection. This may be a self-published book, but there's no hint of any lack of professionalism here. The standard of editing is actually higher than some books I've read from the big five publishers.

It's no surprise, and therefore no spoiler, to reveal that all the different threads of the story turn out to be connected. I soon became aware that the main story was never going to be concluded here. The ending is quite satisfactory, however, as we get to the stage where Detective Thomas has the break he needed in his case, leaving Ford and Mike to continue their quest into the next book in this trilogy.

This is a solid page-turner of a novel, which neatly combines the haunted house sub-genre with noir detective fiction. I certainly have every intention of reading the other two books in the series as soon as I find the time, and I rather hope the story won't end there.

DESCRIBIN' THE SCRIBES

MELANIE ATHERTON ALLEN lives in Pennsylvania with one man, two kittens, and, she darkly suspects, several mice. She could do without the mice, but they seem to give the kittens an interest. Melanie writes horror, mystery, and comedy. Some of her more eccentric projects (for example, an interactive horror story called *The Perils of Sir Reginald*, in which the player tries to beat a family curse before Sir Reginald succumbs to it) can be found on her website, www.athertonsmagicvapour.com.

BRANDON BARROWS is the author of the novels *Strangers' Kingdom*, *Burn Me Out*, and *This Rough Old World*. He has published over seventy stories, selections of which are collected in the books *The Altar In The Hills* and *The Castle-Town Tragedy*. He is an active member of Private Eye Writers of America and International Thriller Writers and was a 2021 Mustang Award finalist. He is also the editor of *Guilty* magazine.

REBECCA BUCHANAN is an author of Pagan poetry, fairy tales, fantasy, mystery, romance, and science fiction, a tree hugger, and a chocoholic. Her collections include *Dame Evergreen* (Sycorax Press), and *The Fox and the Rose, and Other Pagan Faerie Tales* (Asphodel Press), and her essays and short stories are to be found in many venues, including *Sherlock Holmes & the Occult Detectives v1* (Belanger Books).

ROBERT GUFFEY is a lecturer in the Department of English at California State University – Long Beach. His most recent books are *Bela Lugosi and the Monogram Nine*, with Gary D. Rhodes (BearManor Media, 2019) and *Widow of the Amputation and Other Weird Crimes* (Eraserhead Press, 2021). Guffey's previous books include the journalistic memoir *Chameleo: A Strange but True Story of Invisible Spies, Heroin Addiction, and Homeland Security* (OR Books, 2015). He has also written a collection of novellas entitled *Spies & Saucers* (PS Publishing, 2014), and stories and articles for numerous magazines and anthologies.

RHYS HUGHES was born in Wales but has lived in many different countries. His first book was published in 1995 and since that time he has published fifty books and hundreds of short stories. He is currently working on a collection of linked crime fiction stories called *The Reconstruction Club*. His work has been translated into ten languages.

D.G. LADEROUTE is a Canadian author. Their first novel, a young-adult fantasy entitled *Out of Time*, was published by Five Rivers Press in 2013, and was short-listed for an Aurora Award, Canada's premier award for the speculative fiction arts. They also write extensively for the role playing game industry, having contributed to the '*Legend of the Five Rings*', '*7th Sea*' and '*Infinity*' RPGs.

PAUL STJOHN MACKINTOSH is a Scottish poet, writer of weird fiction and translator, primarily based in France. In addition to writing dark tales, he is an active reviewer and critic. He has collections of his fiction and his poetry, a novella, and a ghost-hunting roleplaying game in print. Full details at paulstjohnmackintosh.com

JONATHON MAST lives in Kentucky with his wife and an insanity of children. (A group of children is called an insanity. Trust me.) You can find him at https://jonathonmastauthor.com/

ANDREW NEIL MACLEOD is a Scottish author and musician who lives in Dubai with his wife Amber and their shih tzu, Alex. They can sometimes be found exploring the cobbled streets of Paris, the sandy beaches of Bute where they have a little bothy, or the Highlands of Scotland, where all the magic happens. His novel of Samuel Johnson and James Boswell facing occult menace, *The Fall of the House of Thomas Weir*, is out now from Burning Chair Publishing.

UCHECHUKWU NWAKA is a student of Medicine and Surgery at University of Ibadan, Nigeria. His works have appeared or are forthcoming in *Cossmass Infinities*, *MetaStellar* and *Mythaxis Magazine* among others. When he's not trying to unravel the mysteries of human (or inhuman) interaction, he can be found binging unhealthy amounts of anime, or generally trying to keep up with an endless schoolwork. Find him on Twitter at @uche_cjn.

C.L. RAVEN are identical twins and mistresses of the macabre. They write novels, short stories, comics and film scripts. Their work has been published in magazines and anthologies in the UK, USA and Australia. They've worked on several indie horror films as crew and reluctant actors and have somehow ended up with lead roles in the forthcoming indie horror film *School Hall Slaughter*. In their spare time, they hunt ghosts, host a horror radio show, look after their animal army, and try to look impressive with polefit. Their attempts at gymnastics should never be spoken about.

http://clraven.wordpress.com

CARSTEN SCHMITT is a German SF/F author. He was a finalist for the inaugural George R.R. Martin Terran Award, an alumnus of the 2018 Taos Toolbox writing workshop run by Walter Jon Williams and Nancy Kress, and a winner of the 2021 Deutscher Science Fiction Preis (German Science Fiction Award) for Best Short Story. His stories have been published in Germany, Canada, China, France, Romania, the Ukraine – and now the UK. He lives in Saarbrücken in the southwest of Germany with his partner and three chronically underfed Maine Coon cats.

You can find him online here: www.carstenschmitt.com or on Twitter as @CarsTheElectric

I.A. WATSON stopped actively occult detecting in '09 after the Lochmerle Orphanage tragedy where the black waters rose. Once released from psychiatric care he immediately began scribbling blasphemous truths in the form of novels and short stories and has so far not been restrained again. Amongst his ramblings are a dozen or so novels and collections including *Vinnie De Soth, Jobbing Occultist* (who of course appears in this volume's tale), and more than fifty short stories, a lot of them Sherlock Holmes or occult mysteries for which people keep giving him awards – the unknowing fools! A full list of his literary misdeeds is available at

http://www.chillwater.org.uk/writing/iawatsonhome.htm

CRISTINA L. WHITE is an author and artist who writes plays, fiction, poetry, non-fiction and memoir. Her most recent work is *These Several Women: Love Poems*, a collection of poetry about love, desire, and memory. The collection captures the wonder and beauty experienced when we fall in love; it also speaks to the ache that comes when love takes a turn we didn't expect and everything falls to pieces. The work is inspired by a belief that love is the life blood of our existence here on earth—these poems sing that anthem. She can be found at http://cristinalwhite.com/

Printed in Great Britain
by Amazon

74731232R00133